Frank Hamilton Taylor

Birch Bark from the Adirondacks or from City to Trail

Frank Hamilton Taylor

Birch Bark from the Adirondacks or from City to Trail

ISBN/EAN: 9783337715120

Printed in Europe, USA, Canada, Australia, Japan

Cover: Foto ©berggeist007 / pixelio.de

More available books at **www.hansebooks.com**

Birch Bark from the Adirondacks;

OR,

FROM CITY TO TRAIL.

By FRANK H. TAYLOR.

ISSUED BY THE ADIRONDACK RAILWAY COMPANY.

ILLUSTRATED AND PRINTED BY THE
LIBERTY PRINTING COMPANY,
107 LIBERTY ST., NEW YORK.

Season 1887.

Jaques.—"Which is he that killed the deer?"

First Lord.—"Sir, it was I."

Jaques.—"Let's present him to the Duke, like a Roman conqueror; and it would do well to set the deer's horns upon his head, for a branch of victory. Have you no song, forester, for this purpose?'

Second Lord.—"Yes, sir."

Jaques.—"Sing it! 'tis no matter how it be in tune, so it make noise enough."

SONG.

What shall he have who killed the deer?
His leather skin and horns to wear.
Chorus—Then sing him home.

Take thou no scorn to wear the horn;
It was a crest ere thou wast born:
My father's father wore it.
And thy father bore it:
The horn, the horn, the lusty horn,
Is not a thing to laugh to scorn.

"As You Like It."

HE ownership of a gun is an expensive responsibility—that is, of a *good* gun. The original investment is nothing compared to what follows.

You see, when a man has a nice double-barreled fowling piece, breech-loading, of course, and with all the essential attachments made and provided; or a Winchester rifle, with a belt full of cartridges, forty-four calibre, he is very sure to think of them some hot summer-day when he ought to be concentrating his mind on the chances of fall trade, and just as soon as he begins to think of his gun, he is certain to become reckless, and, instead of withstanding evil temptation, sets himself right in its way, and we may look for him and his wife and youngsters upon the deck of one of the big and splendid and fast steamers of the "Citizens'" or "People's" line, or of the "Day Line," with tickets for Saratoga and beyond; or the next morning, or the next after that, he puts his family upon the Saratoga train at the Grand Central Depot, rides with them all the forenoon along the changeful scenery of the lower Hudson, reaches Saratoga, turns his back upon its manifold allurements and vanities, transfers Mother and Emily and Jim and Dot, the baby, with their several and particular bundles, valises, *et al.*, aboard the train of the Adirondack Railway, and goes on northward to Luzerne, Warrensburg, Chestertown, Riverside, Schroon Lake, North Creek, Blue Mountain Lake, Raquette Lake, Long Lake, or the vast and silent wild-wood solitudes beyond. If he halts at Luzerne he is there presto! but if to the forest-environed lakes his adventurous family is swallowed up in the depths of a stage-coach, or safely disposed upon its summit, or tucked upon a "buckboard," and there they stay until the wheels have measured their daily task along shadowy brooks, over rugged hlils, through cool forest avenues and once in awhile past spasmodic evidences of agriculture. Now and then a little rude wayside hotel invites a stop and the thirsty horses are watered. The stage ride from Riverside to Schroon Lake is but the matter of an hour, while the winding but excellent highway to Blue Mountain Lake is thirty miles long; few, however, tire of this well-kept woodland avenue or its varied scenery. It is worth while to remember that the new and admirably equipped stage line, organized in connection with the Adirondack Railway Company, is known as the Blue Mountain Lake Stage and Transportation Company, the only line over which through tickets are sold from the large cities direct to Blue Mountain Lake, and over which baggage is checked through.

Each season new efforts are made increasing the attractions of the region traversed or reached by the Adirondack Railway. This year "The Adirondack, Lake George and Saratoga Telegraph Company" has been organized, and owns and operates the various telegraph lines between Saratoga, Luzerne, Schroon Lake, North Creek, Blue Mt. and Long Lakes and points beyond in the Adirondack Mountains, giving better service and much lower rates than under the old system of numerous independent lines.

Arrangements for a direct stage to Long Lake via Blue Mt. Lake are to be accomplished facts this season, money having been appropriated to improve the road between Blue Mt. and Long Lakes.

If our friend of the gun goes to Schroon Lake he has a steamboat voyage as a final variation of his journey; or if he takes the longer stage journey westward the sunset falls in mellow radiance upon the shapely flanks of Blue

Mountain and its sleeping lake purples in the dying light when he and his brood dismount at the Lake House, Blue Mountain House, or the large and modern "Prospect."

At either of these lakes he will find scores of other fellows who have also been tempted by too much contemplation of their guns or fishing tackle into deserting their several customary occupations and rushing off with their families or especial friends into the Adirondack woods.

The intending traveler who does not mind the hum of the mosquito and is indifferent to its sting, who fails to find the bite of the "blackfly" an annoyance, rather enjoys oil of tar, and only hankers after fishing, such a one may penetrate into the woods as early as May 1st; but the man with a family, the man caring for comfort and repose, and seeking that rest to body and mind which the Adirondack life and the pure balm-laden air will surely give him, should arrange his arrival and time his stay between the early part of July and the middle of October, for during these months the insect pests are harmless and the course of life is joyous, and the cobwebs are blown away from tired brains, and fresh vigor enters into wearied bodies.

If all these nimrods and Waltons come here to fish or blaze away at small promiscuous game they may begin the slaughter almost any time. The trout are "at home" after May 1st, but the law makes it expensive to kill a deer before August 1st, or after November 15th. There are *some* people who think that even if a man don't come here to shoot a deer or catch a three-pound brook trout, there is enough reward for coming in the simple pleasure of breathing the pure air, voyaging in the light Adirondack boats from lake to lake, following the winding trails,

AT LAKE LUZERNE.

watching the sun come up from behind Blue Mountain and go down into the midst of the western forest, and so everybody is satisfied, and the city man's conscience stops troubling him for neglecting his affairs, and somehow he begins to think that the investment in his gun was a pretty good thing after all. When he has listened for a week or so to his guide's fund of reminiscence and imbibes something of the flavor of the simple, healthful life he finds amid these upland waters he begins to wish he might stay here every month in the year—except June.

When our truant goes back to his work in the city, as brown as a pirate and feels once more the tyranny of the linen collar about his neck, he finds that, after all, the world has jogged along about as usual, and that the fresh grip he can take upon his own affairs more than compensates for the loss of a few possible bargains or turns he has missed in the market. Having learned this great lesson in the practical philosophy, he not only persists in rejoicing in the fact that he owns a gun or a rod, but straightway wants all his friends to get one or both. He also wants to tell them all about the Adirondack Lake Region, the finest "preserve" east of the Rockies, and as the memory of what happened last year is apt to become hazy, say as to the distances, size of the trout, plentitude of the deer, and the like, I will begin my litttle book with the memoranda which is finally collected from my several pockets, while I sit by the fire after a spring day's journey, which will, I hope, help to preserve him and all travelers either from the shoals of under-statement or the quagmire of exaggeration.

For more specific information the reader is referred to "Wallace's Guide to the Adirondacks," Syracuse ; "Stoddard's Adirondack's," Illustrated, Glens Falls ; and "Stoddard's Map of the Adirondacks."

These may be had at Saratoga Springs, and at the news stands and hotels through the region of which they treat.

Each of the authors of the above works has traveled for years in these woods, and has labored *con amore* to make the paths through the wilderness plain to the visitor.

NORTHWARD FROM SARATOGA.

The Adirondack Railway has its beginning at Saratoga, its trains leaving the same depot at which passengers upon the "Delaware and Hudson" trains alight. It extends fifty-eight miles slightly west of north to North Creek, the grading for some miles further giving promise of its extension beyond, at no distant day. Leaving the shady avenues of Saratoga behind, the traveler will note the fine grounds of Hilton Park upon the right. Toward the westward there is a beautiful vista of cultivated lands and vine embowered homes. Saratoga Lake gleams fitfully through the verdure miles away to the southward as we take a sunset course.

At the station of Jessup's Landing a stage awaits which connects with the village of Corinth, one mile distant, and Palmer's Falls, two-and-a-half miles away. The extensive manufactory at the latter place of the Hudson River Pulp and Paper Company and the immense water power

9

still unused justify a branch railroad which will not only facilitate the heavy freightage but will enable passing tourists to view one of the most picturesque cascades in the Empire State. This branch road will be completed during the present season.

Here the entire outflow of the upper Hudson pours down in a wild delirium over the rocks, giving an enormous water power which will, at no distant day, turn the wheels of a group of vast industries.

The large establishment already located here uses 10,000 horse power in the work now operated, the product being about ten car loads per diem of wood pulp or paper, all of which is consumed by leading journals. The wood used is spruce and popple, and the process of reducing the solid logs to a creamy pulp and then forming the pulp into smooth and crisp sheets of snowy paper ready for its daily burden of news for good and evil, is very interesting. The works employ about 250 men, and the present capacity will shortly be doubled. The daily product of paper alone would form a single sheet not less than two hundred miles long. The officers of the company are: R. Pagenstecher, President; Hon. Warner Miller, Secretary; A. Pagenstecher, Treasurer and Warren Curtis, Manager and Superintendent. The capital stock is $240,000, and the stock is now estimated at twice par value.

Fifteen miles from Saratoga the Hudson River is met again coming down from the highlands we are invading. There is a bright play of color along its banks in the glades of deciduous trees still bearing their tender early greens and reds, with sombre pines and cedars for a background. Imposing groups of these noble pines are frequent along the river, and serve as a reminder of the days when this whole region was covered with these great trees, and the moccasined Indian trod softly the mossy carpets that enfolded their roots.

Just before reaching Hadley, the station for Luzerne, twenty-two miles upon our way, the railroad crosses on a fine bridge, ninety-six feet, above the water, the Sacandaga, a tributary of the Hudson, into which it pours a foaming, cedar-stained flood. You leave the cars at Hadley, and passing on your right the pulp mill of the Hon. George West, cross the Hudson to Luzerne, a half mile from the depot. Luzerne is sacred in the memory of many who tarry at Saratoga because of its delicious trout dinners. There are few villages among our inland hills more healthfully located than this little namesake of Switzerland's famous resort. Its resident population forms a delightful social nucleus reinforced during the summer by many well known city families. Its proximity to Saratoga enables residents to enjoy the life at our great American Spa. Special trains will be run this season between these points at hours which will enable visitors at either Luzerne or Saratoga to make rapid and frequent excursions to and fro. There are many agreeable and picturesque drives in the vicinity. Opposite the village the Hudson pours down over a rocky ledge forming a picturesque cascade. A tributary having its source in the Palmerton range is caught above the town and forms the bright and placid Lake Luzerne. The hotels — Rockwell's and the Wayside, are filled all summer with city folks, and there are many private boarding houses. Rockwell's is in the heart

BRIDGE OVER THE SACANDAGA AT HADLEY.

THE UPPER HUDSON.

of the village, and hardly needs an introduction to the experienced traveler. Capacity, about 150. The Wayside faces pretty little Lake Luzerne and with its cottages accommodates 250. The distance from this point to the Fort William Henry Hotel, upon Lake George, is about ten miles.

Lake Luzerne is an irregular sheet nearly surrounded by steep hills. The Lake itself and the "Creek" running into it are charming retreats, cool and inviting, and the shady paths around them tempt the sojourner to the pleasures of pedestrianism. There is a choice of pretty churches for the Sunday services, and nowhere is the social status of summer visitors higher, or the intercourse more agreeable than here.

AT THE OUTLET—SCHROON LAKE.

Nothward from Hadley and Luzerne the railroad keeps along the upper Hudson which here looks, with its log-burdened surface, very like the west branch of the Susquehanna near Lock Haven, or the James, in Virginia, near Lynchburg, or a dozen other young rivers I have in mind which flow fretfully down their rocky beds from wild timber countries where the axeman is busy. Potash-Kettle Mountain interposes a barrier as difficult as its name between us and Lake George.

There are stages now and then waiting the certain mail-bag and probable passengers at the little stations, these leading away over lateral roads to pleasant out-of-the-way corners of the world which some summer travelers have found and return to enjoy.

There is one at Hadley which will take you, if you wish, up the Sacandaga Valley to Conklingville, another at Riverside for Schroon Lake; and lastly the roomy vehicle standing beside the platform at North Creek which is to bear us off, if we so choose, to the charms of Blue Mountain Lake and its limpid waters; and still another to Long Lake or to Adirondack Village, upon the sunset side of Mount Marcy, where the wonderful Iron Ore beds are situated, of which State Geologist Prof. Emmons has written so much, and which, as soon as the Adirondack Railway is extended towards Long Lake, will turn out hundreds of thousands of tons of ore yearly, for the manufacture of steel and iron. It is near these ore mountains that the Adirondack Club has its headquarters.

At present we are bound for Schroon Lake.

THE SCHROON LAKE REGION.

Seven miles by stage from Riverside brings the tourist to Pottersville, where meals *en route* are usually taken at the hotel of the same name. A short ride beyond reveals the outlet and steamer landing. The balance of the journey to the village of Schroon Lake is made upon the steamer "Effingham."

At Riverside, the shady camp ground of the Methodists comes into view upon the opposite side of the river. At this point, too, the stages for Schroon Lake meet passengers destined for the several hotels upon that picturesque sheet. The vehicles in use are well-made and roomy Concord tally-hos, and after the more conventional journey by steam, the seven miles of merry wheeling over a good road to Pottersville, is a welcome diversion. The road leads through shadowy pine groves and along the margin of Loon Lake and one or two pretty ponds in whose still waters is reflected back the bright overreaching foliage of their shores.

Passengers arriving at Riverside upon the early morning sleeper, breakfast at Pottersville, at the hotel of that name, and those coming later in the forenoon dine there as do parties returning from the lake. This house is owned by Messrs. L. R. & E. D. Locke, who are also proprietors of the Leland House and cottages at Schroon Lake village. The stage company carries passengers, after meals, to the landing at the outlet, a short distance beyond the village. Here the steamboat "Effingham" receives the tourists and proceeds up the lake, touching at intermediate landings, to the village at the northern extreme, where most of the hotels are located.

Schroon Lake is long, narrow and crooked. Throughout its length it is environed by noble hills, and from its many rounded headlands a great variety of pleasing outlooks greet the visitor's eye. It bears a resemblance to the lovely lakes of central New York, especially Otsego, while its surroundings are far more wild and imposing; indeed, the vistas are suggestive of the vaunted and much-painted Lake George, from which it is separated by twenty miles of wilderness.

The excellence of the passenger service both by rail and stage has done much to tempt city people of wealth and taste to come to Schroon year after year, and the pretty cottages many of them have built peep out from the foliage as we speed along the smooth blue lake.

Three miles from the outlet the lake attains its greatest width, and here, upon a shapely knoll, reaching out into the lake, stands the Taylor House with its group of a dozen neat cottages, half concealed—like a matronly old hen and her chickens, among the fragrant pines and balsams.

Although the Taylor House has been built but a few years it is a favorite with many old habitués, as well as more recent visitors, both from the beauty of its location and the unfailing excellence of its fare. The name of this place, Lake View Point, will suggest the far-reaching panorama of water and upland which pleases the eye both up and down the lake.

SCHROON LAKE, FROM THE TAYLOR HOUSE.   LAKE VIEW POINT.

The resources of a well-stocked farm of two hundred acres are drawn upon to supply the table. Cool, pure spring water, brought from the mountains, bubbles up through the grounds, and is carried by pipes into every structure, the elevation of hotel and cottages above the lake making drainage perfect. Plenty of boats and a good livery are among the attractions of the place. One of the finest drives I have seen among the lake regions of the Empire State leads along the western margin of the lake to Schroon, and thence into the still wilder regions to the northward.

The Taylor House maintains a telegraph and a post office, and good instrumental music for dancing and promenades. Mr. C. F. Taylor enjoys the distinction of being the pioneer hotel man upon the lake, and he is now prepared to care for about two hundred guests.

Just beyond Lake View Point the large and handsome summer home of Edward Harrigan, the favorite comedian, is located.

Opposite, a mile and a half across the shining lake, is the village of Adirondack, nestled in a dimple between the hills. Here Mill Brook pours down from the rocky fastnesses above and a large tannery is located. The Wells House at this place, recently rebuilt, accommodates about 50 guests, and is a most excellent house. A good road along the lake connects it with Pottersville, vehicles meeting the stage at that point. Midway up the lake to Adirondack is the Granger House upon this road. The highlands opposing approach each other and through them to the right we catch a first glimpse of Schroon Village, its large hotels showing brightly in front.

Schroon Village faces almost directly toward the south, and with the cordon of lofty mountains in its rear it enjoys an excellent and moderate temperature which should materially prolong its "season."

The pine clad Isolabella, occupied by the family of the late Col. Clark, is upon the right. Directly ahead is the handsome front of the Leland House with its flanking cottages and sloping lawn dotted with brightly costumed guests; to the left is the well-known Lake House and in the pleasant village behind them are the Windsor, Ondawa and Arlington, all familiar to visitors of former seasons.

About a mile from the village upon the western shore is Bishops, a boarding house, close beside which is a small but very ornate building used by guests as a club-house.

The Leland House at Schroon Lake Village is located upon the rising lake shore at a point which commands a view of nearly the entire lake. Money has been lavished upon the adornment of the surrounding grounds, which include a fine tennis level. The hotel and its adjacent cottages can house about two hundred persons, and during the season the names of many prominent citizens are inscribed upon its register. As already stated this favorite house is conducted by Messrs. L. R. & E. D. Locke of Pottersville. The grounds of the Lake House adjoin those of the Leland and are also upon the lake front. This house has a capacity of about 100, and is well filled during the warm season. The Ondawa, capacity about 80, is open through the year. The Windsor, also near by, can entertain about 100. The Fowler House,

SCHROON LAKE.—1. A VILLAGE CHURCH.　　2. THE VILLAGE, FROM THE LAKE.　　3. ISOLABELLA.

north of the village, Leland Cottage, Russell Cottage, Hazelton Cottage and Grove Point House, can each take about 30 visitors.

The important question of fishing in the lake can be answered to the satisfaction of anglers. The lake has been abundantly stocked with the gamey lake trout of which many large specimens are caught each season. Black bass and pickerel abound. In ponds, streams and smaller lakes, near by, the brook trout are plentiful and lively. Mr. C. E. Benedict has a successful trout hatchery at his handsome place upon the road along the lake a couple of miles from Pottersville, from which streams and trout waters in the vicinity are supplied with young fish.

A point which will be appreciated by summer visitors is the fact that the New York morning papers are taken up the lake upon the boat leaving Pottersville about 3 p.m.

There is much driving at Schroon Lake, the favorite routes being to Paradox Lake—where, as I hear, marvelously appetizing trout dinners are served—and to forest environed Pyramid Lake.

There is an outlet for travel toward the north via the Elizabethtown stage, which leaves upon alternate days, the distance being 32 miles.

Root's old staging house, one of the first in the mountains, is nine miles north of Schroon Lake. Throughout the Schroon region deer are plentiful, as well as partridge, quail and other small game.

The wild and picturesque Champlain slope of the Adirondacks to the north and east, is peculiarly within the realm of the Delaware & Hudson Railroad Company, the great trunk line between Montreal and New York, which connects with all the highways into the eastern and northern Adirondacks.

From Riverside station, where we left the railway, a stage goes to the pleasant village of Chestertown, a distance of about six miles. The Chester hotel has a large summer patronage. Mr. G. W. Ferris is the proprietor. The Rising House at Chestertown is well known to many travellers in this region. Summer boarders are also taken at the farm house of Royal P. Mead. Schroon, Brant, Friend and Loon Lakes are all within five miles of Chestertown, and are reached by pleasant and winding roads. Lake Pharaoh affords the best trout in Northern New York. The ponds and brooks in the vicinity are full of that most delicious fish, the brook trout, and the sport will repay the angler's skill.

An omnibus will run from the Chester House to all the lakes in the vicinity at twenty-five cents a passenger where there are parties of six or more. It will remain with the guests to make up a party each day if they wish. Quiet and comfortable boarding houses are found at these latter places, and although there is no startling scenery, no big lakes or rivers in the vicinity of Chestertown, yet those caring for solid comfort, pure air and a pleasing landscape; with pretty drives and all at moderate figures, will do well to remember this easily accessible neighborhood.

ACROSS THE DIVIDE.

The thirty-mile stage route from North Creek across the Divide, separating the upper waters of the Hudson from the watershed of the Raquette, has been the recipient of much attention at the hands of several road gangs the past and present seasons. With a large outlay of money, intelligently applied, it has been vastly improved by the Adirondack Railway Company, and owing to the enterprise of the owner of the Prospect House, Blue Mountain Lake, three miles of the western portion has been newly located and turn-piked.

A pleasant feature of travel over this highway is found in the excellent "buckboards" used, which many travelers prefer to the stage. These "buckboards" are really elegant carriages, and, unlike those so popular at Mount Desert, they have tops and side curtains. Next to the Cuban volanté they are the most comfortable vehicles possible in a hilly region. They gain their name from the use of the "buckboard" wooden spring combined with regular carriage springs of steel.

This road, after leaving the terminus of the railway, keeps along the diminished Hudson as far as North River, about five miles. Here is Eldridge's, a pleasant little wayside house, which can furnish a most palatable meal; and where tourists, desiring a quiet retreat, will find pleasant quarters at reasonable charges.

From this point the road climbs to a higher plateau, coursing through a wild and rocky

BLUE MOUNTAIN.

section. Be sure to ask the drivers to point out Mount Marcy, on whose slopes the Hudson River takes its rise. The way leads through the settlement of Indian Lake with now and then a more or less rude, but hospitable inn, where, beyond a doubt, a surprisingly good meal could be speedily conjured if we but asked it. With good fortune, however, we mean to sup at Blue Mountain Lake, and with a half regret that the pleasant ride is not longer, we finally bring up before our hotel and submit our travel-stained visages to the inspection of the maids and matrons there congregated.

This mountain air *does* give one an extraordinary power for assimilating food, and somewhat tired with a long day on wheels we turn in early.

Now in the bright morning light let's take a look at our stopping place, the Prospect House.

The main structure of five stories and a length of 255 feet fronts upon the lake, from which it is separated by a lawn broad enough for a brigade review. Blue Mountain, which *does* seem a bit more ultramarine than its sister heights, is opposite in plain view.

From this main building a wing of 150 feet projects back toward the road. A grand piazza of extraordinary width extends along the entire front and ends, with a series of detached balconies above. Pointed turrets give relief to the centre and corners of the roof. This fine structure contains 200 large, airy and pleasant rooms, in addition to the spacious office and general room, dining hall and parlors.

PROSPECT HOUSE, BLUE MOUNTAIN LAKE.

This was the first hotel in the world to introduce the Edison electric lights into sleeping apartments. The hotel is supplemented by a handsome cottage, with open fire-places in all rooms, which is kept open throughout the winter. Running water, a steam elevator and all other conveniences that fastidious travel demands will be found here.

A billiard room, bowling alley and shooting range occupy a separate structure. Lawn tennis grounds are provided, and an extensive fleet of skiffs is ready at the wharf. Telegrams may be sent from the house to any part of the world. A handsome windmill, set close to the new sea-wall along the lake, is a recent and prominent addition to the property.

The Lake House, a handsome four-story building, rebuilt and enlarged since the destructive fire of last November, is honored as the pioneer hotel upon the lake, the date of its erection being 1874, and Mr. John Holland, its proprietor, is the first landlord who made a visit to the lake a pleasure to travelers by a welcome to the hospitable log structure which was first used as a hotel upon the same site. It is the first structure seen upon approaching the lake, and stands upon an elevation commanding a fine view. Ten cottages are connected with the hotel, giving a capacity for entertainment of about two hundred.

CAPT. HANK BRADLEY.

27

MOONLIGHT IN THE WILDERNESS.

It is well patronized, and has a capital lawn tennis ground and fine sand beach in front. The Episcopal log church is near this hotel.

The Blue Mountain House stands upon a lofty site upon the slope of the mountain. A good carriage road connects the house with the valley, and stages call upon regular trips. The former house which was destroyed by fire six years ago was replaced by a larger and more modern house, having a capacity of about sixty. The proprietor is Mr. Tyler M. Merwin. At this hotel the charges are very moderate, and persons of slender means can safely patronize it.

Blue Mountain has an altitude of 3,824 feet. A good bridle path has been completed to the summit, and a saddle livery will be maintained this season, for the convenience of excursionists who wish to gain the superb view from the summit.

The altitude of Blue Mountain Lake is nearly 2,000 feet. It is one of the loveliest of these Adirondack Trossachs.

Forest-clad islets and points are occupied by handsome cottages, which suggest the Thousand Islands. Messrs. Duryea of New York, Thacher of Albany and Crane of Yonkers having the most attractive on this lake.

Owing to a recent change of ownership of Blue Mountain, Eagle and Utowana Lakes and the lands about and around them, it is confidently believed that persons desiring to purchase cottage and camp sites will now be able to do so to a limited extent. More definite information concerning this matter can readily be obtained of The Adirondack Railway Co., 20 Nassau St., New York.

GUIDES AND BOATS.

Beside the attractions of the hotel, with its sparkling kaleidoscopic summer life and the wild wood beauty of its immediate environment, there are two important factors to the pleasure of the visitor, and these are inseparable.

They are the guide and his Adirondack boat.

The guide without the boat would be as a musician minus his instrument. While the boat, deprived of the strong arms of the woodsman to pull the oar, his broad back to bear it across the toilsome "carry" and his intimate acquaintance with every hidden reef, impracticable rapid or false trail, would be useless indeed. There is probably no craft in the entire marine category lighter for the work it has to do, than an Adirondack boat, and yet in the hands of the man who takes you through the scores of lakes and leagues of little rivers it is as safe as a ship.

The Adirondack guide is nearly always a native of the woods and hills. There is a wide contrast between the highlander and the lowlander whenever you find him. I know the mountaineers of the Appalachians from Memphremagog to Georgia and I know of no finer, more manly race. They are, with the unerring instinct of the forester, good judges of human nature. Their independence of bearing is their best heritage. The study of the woods and waters, the haunts of the stag and lurking places of the trout is their capital, and when we hire our man and his boat we get far more than at first appears. As he pulls away through the chosen tour he opens up to us a vast fund of woodland lore. His stories, though every squirrel in the woods may have heard them to the echo, are new to us. He can broil a trout or flirt with brown lake-side maids with equal facility. His good nature is the infallible outcome of a perfect digestion, and while he finds dollars plentiful but a short portion of the year, an unprejudiced outsider must find the conviction forced home upon him that here is a man who is living closer to the methods and life for which nature intended human beings than any of us who are tossed upon the social seas of city existence. The Adirondack boat has gained a wide reputation, and it is probably the most capacious craft for its size and "heft" yet devised. The average weight is 70 lbs. The bow and stern lines are exactly alike, the cutwater projecting at the water line with a graceful ram-like curve. Its profile is not unlike the aboriginal birch-bark. It is made of thin pine strips lap-streaked and copper-fastened. The ribs are set quite closely and are sawn from carefully selected roots of the same general curve.

The oarsman sits closely into the bow and his passenger equally near the stern. The oars are small, light, and stroke short, with a troublesome lap to the handles that "breaks up" an oarsman accustomed to the broad sweep of the St. Lawrence oar, but a little practice soon overcomes this peculiarity.

A yoke similar to those used in the sugar-bush is strapped with buck-thongs to the thwart amidship, which supports the weight of the boat upon the shoulders at a "carry." Fashion (or possibly a law with due penalties, for aught I know) has decreed that all Adirondack boats should be painted dark blue outside and green within. The average distance made in a tour of a week of continuous travel in such a craft will be twenty-five miles per diem.

Now you are introduced to your guide and your boat, let me inform you that both await you at the landing, while you are saying "good-bye" for the fifth or tenth time and nervously overhauling your *impedimenta* in dread of forgetting something. You'll wish you *had* forgotten about half of that *plunder* before you get across your first "carry" beyond Forked Lake.

The initiatory miles of your tour bear little evidence of the "roughing" which you have been led to expect.

The "Toowahloondah," a neat steam yacht, not quite so long as its name, bears you and your companions, with your respective guides and boats, away swiftly and joyously down the lake.

Now, let's take a look at our excellent maps. This is Blue Mountain Lake: this little spot just on the edge on the lake region. As you look along its blue and breezy expanse to its distant shores you imbibe some notion of the extent of this great wild park of the Empire State.

EAGLE LAKE OUTLET.

FOREST MONARCHS.

Out of Blue Mountain into Eagle through a narrow and winding channel we go. Eagle Lake is quickly crossed, and then comes the longer pond of Utowana. But we must not forget to notice, nor will our guides, let us do so, the hill-side log house and barns, where Col. E. Z. C. Judson, otherwise Ned Buntline, spent a portion of his life and wrote much of the romantic fiction with which his name is associated. After a varied and exciting career, the author and poet met his end quietly at Stamford, at the source of the Delaware, in the Shandaken Catskills. These were the stirring lines which we read from Stoddard's book, as we traversed the lake the hunter, novelist and poet loved to glorify:

" THE EAGLE'S NEST.'

Where the silvery gleam of the rushing stream,
Is so brightly seen on the rocks dark green,
Where the white pink grows by the wild red rose
And the blue bird sings till the welkin rings.
* * * * * ⁂

Where the rolling surf leaves the emerald turf,
Where the trout leaps high at the hovering fly,
Where the sportive fawn crops the soft green lawn,
And the crow's shrill cry bodes a tempast nigh—
There is my home—my wildwood home.

At the western extreme of Utowana we come to the upper end of the Marion River carry. If you have your rod ready, early in the season, you can probably yank a two pound trout out of the little outlet rapid below the dam while the wagon is being laden with your "fixin's." This is only 'a probability, however, bear that in mind.

We trudge away over the half mile wagon road leading to the landing on Marion River.

"After all," you say, "this 'carrying' I've heard about aint very bad; pretty good road wagon to 'tote the duffel,' as they say up here." "Now, just you wait, mister; that ere wagon aint goin' along with us, nor this boulevard, neither," replies your guide, from the recesses of his boat.

At Marion River we meet the steam yacht "Killoquah," a trifle larger than its sister we have just abandoned.

The same sturdy mariner who has brought us thus far, Capt. Hank Bradley, takes the wheel. He can't talk to you now, but when he reaches big water just ask him about his trip on an ice-boat on Raquette Lake.

If there's anywhere in the Kingdom another river as completely erratic and purposeless in its wanderings through a marsh as the Marion, I only know of one, and that's the Ocklawaha in Florida. We are dizzy when we reach the opening looking out across the Raquette.

While voyaging around the many headlands of this famous lake, one is at once struck with its great similarity with Winnepesaukee, in New Hampshire, and Minnetonka, near Minneapolis. Like those favorite resorts it is an aggregation of bays and islands. It is full of charming surprises, and a very paradise for the canoeist.

The chance voyager will be tempted to halt for a day or so at a picturesque hotel seen to the left soon after leaving the Marion. It is flanked by a couple of ornate log cottages. The whole is called "Under the Hemlocks." The proprietor is Edward Bennett.

THE "KILLOQUAH."

AFLOAT AND ASHORE.

In addition to "Under the Hemlocks," the tourist will be made welcome at Hathorn's Forest Cottages at Golden Beach, South Bay. The structures are of solid logs, with all the charming woodsey air that is so attractive to all who love out-of-door life in the woods. Stoddard's excellent photographs, to be found everywhere in the mountains, will give a good idea of this charming place. The proprietor is Chauncey Hathorn.

Isaac Kenwill welcomes the voyager at the Kenwill House. This place has a capacity of about forty, and is favorably located upon a charming peninsula. Mr. Kenwill, himself, has long been known as an experienced woodsman, while his wife is an admirable hostess.

Blanchard's Wigwams accommodate about twenty. Boarders are also taken at Joe Whitney's Camp on South Bay.

A new boarding cottage, and a series of rustic lodges and dining-rooms, under the general title of "The Antlers," will be opened this season by Charles H. Bennett, on Sand ; or, as it is also called, Constable Point ; where a delicious spring of the purest water bubbles out of the forest glade, and whence a magnificent view of the lake and mountains is to be had. It is plainly in view after leaving the Marion River upon the western shore of the lake. The Post Office adress of all the Raquette hotels is Blue Mt. Lake, Hamilton Co., N. Y.

Happy indeed is the adventurer into these wild-wood scenes who has an "invite" to a Raquette "camp." Was there ever a word so misused?

Filled with preconceived notions or rude and temporary shelters and culinary makeshifts

37

CAMP STOTT AND CAMP FAIRVIEW, RAQUETTE LAKE.

encountered in other woodlands, this neophyte in Adirondack experience comes upon a shapely villa of solid logs, set beneath the grand old hemlocks and pines, with accessory buildings for cooking, dining, sleeping, the children, the guides, and what not—a village in rustic, touched with an elaboration of interior furnishing which tell not so much of wealth as of loving care and exquisite taste.

It does not cost much to build a bedstead of cedars with the bark preserved, but it requires taste to conceive the idea. And the chimneys—the very rallying points of hospitable intention—wide-mouthed, massive stone affairs, sheathed to the ceiling with bark and

AN INTERIOR AT "THE CEDARS."

buttressed with saplings; firey shrines dedicated to the gods who bring us good cheer and comradeship. Camps, indeed .

Somewhere along the rocky shore, with an eye to picturesque effect, you will find the " open camp," which is an institution peculiar to the Adirondacks.

"THE CEDARS."   FORKED LAKE.   CAMP PINE KNOT, RAQUETTE LAKE.

AN INTERIOR AT CAMP PINE KNOT.

41

It is a shed of logs and bark, lined with birch bark and heaped within, a foot or so deep with fragrant cedar boughs.

This is the story-teller's haunt o' nights, when firey tongues creep up amid the pyre of logs in front, and the shadows play fantastic tag with the red light among the silent hemlock boughs o'erhead.

There are several of these "camps" upon Raquette Lake. They are owned as follows:

Echo Camp, on Long Point—A. S. Apgar, of New York; Camp Fair View, on Osprey Island—C. W. Durant, of New York; Camp Pine Knot, on Long Point—W. W. Durant, of Saratoga Springs; Camp Stott, on Bluff Point—Frank H. Stott, of Stottville, N. Y.; Teneyck Camp—James Teneyck, of Albany, N. Y.; McCarthy Cottage, on Kenwill's Point—Mrs. Senator McCarthy, of Syracuse; Camp Hasbrouck—Mr. Hasbrouck, of New York.

Culture and the intention of permanency has brought with that recognition of the Supreme and Beneficent Power which reigns in the forest and the city alike. This has taken shape in the form of a beautiful little church and rectory set upon an islet, like a lighthouse upon this remote frontier of civilization, that he who will may worship.

Regular service is conducted here throughout the summer, the attendants at the church arriving and departing in boats.

The "Killoquah" will carry the voyager and his retinue to the further extreme of the lake, and set him down at another carry near the outlet. You make up your mind about how much

GOING TO CHURCH—RAQUETTE LAKE.

you can lug for half a mile and tackle it. The guides have already disappeared in the leafy distance. They will come back and do your "carrying," if you want them to do so, with the utmost good nature, but both economy of time and self-respect urge that you should not go forward empty-handed unless you are an invalid.

A plain but neat building at the end of the carry is known upon the "act for the protection of innkeepers, etc.," nailed to the chamber doors as the Forked Lake House, but popularly as Fletcher's. The good fare and fragrant coffee you may get at this point upon your tour will be recalled more than once before you return.

Skirting the southern margin of Forked Lake to its outlet eastward brings us to a long wildwood "carry," which is a fair specimen of those we are to encounter thereafter.

"This is the forest primeval." How those stately lines of Longfellow's recur to the mind as we pause in the deep midst of the trail to catch our breath and rest our burdened arms. This is the very home of silence and the little noises our small procession makes seem a profanation in the aisles of this temple of nature, but we trudge on unheedingly, devoting our best energies to the effort of keeping apace with the swift-footed guides.

There is the deep diapason of falling water ahead, and shortly we are abreast the Buttermilk Falls, a snow-white cascade, below which our Adirondacks are floated, and we speed down the swift little river into Long Lake, after another short but lively "carry."

The vista of Long Lake is pleasing from the start. Mt. Seward rises in the north with lesser hills in front, across which the cloud shadows drift and give them shape. Midway up the lake a roomy hotel, the Sagamore, looks out toward a round and wooded island. The Sagamore is the largest and newest of the several hotels upon this fine lake. A floating bridge traverses the lake at this point, along which a solitary youth, suggestive of Whittier's "Barefoot Boy," is busy with a dozen poles set for pickerel. Some genius has, within a few years, introduced this

BUTTERMILK FALLS.

voracious and destructive fish into this and its connecting lakes. No self-respecting trout will occupy the same territory with this fresh-water Ishmaelite; and you needn't try your fly upon Long Lake, though the tributary brooks offer better luck. Long Lake Village, hidden behind a ridge and half a mile back from the lake, has a neat little hotel, much liked by those who have summered here. There are two or three well-stocked stores.

A useful specialty of Long Lake Village is its Indian portage baskets, which are to be strapped to the shoulders like a knapsack and will hold a surprising quantity of "duffel." These baskets are narrow-mouthed and deep, and as whatever you happen to want is very sure to be at the bottom, they offer a pleasant and frequent means of occupying the attention in transit. Here also are built in perfection the famous Adirondack boats.

Forked Lake

Round Island, Long Lake

Long Lake, Mt. Seward

AMONG THE LAKES.

Just here is a point for you. A daily stage navigates between Long Lake Village and Blue Mountain Lake during the season. The distance is only eight miles. Thus you can arrange to have supplies or extra baggage sent over to you, or what is more likely, have some of your load sent back to your anxious friends, or the friends themselves can take a little ride over to meet and condole with you. Again, if your tour is limited in time, and you have already seen Raquette and Forked, you can save a couple of days and take your boats at this place.

The Sagamore, the large hotel already mentioned, accommodates 150, is newly and elegantly furnished, and the rates are $12.00 to $20.00 per week, or $3.00 per day. Address E. Butler.

The Lake House, Mrs. C. H. Kellogg, has a capacity of 50. Rates, $10.50 per week ; $2.00 per diem. Grove House, David Helms, holds 30. Rates, $8.00 to $10.00 per week ; $2.00 per diem. Long Lake House, Helms & Smith, capacity 20. Rates, $10.00 to $12.00 per week ; $2.00 per day. Austen's Cottage, Henry Austen, accommodates 15. Terms, $7.00 to $10.00 per week ; $1.75 per day.

The Island House, near the foot of the lake, is managed by Wm. Kellogg, accommodates 12. Rates, $10.50 per week, or $2.00 per day.

This particular portion of the world does not exactly flow with milk and honey, but there *is* a marked exudation of maple sap and spruce gum, the latter being another active specialty. Its use is widespread among the natives, as it is considered a great promoter of appetite. Now, one would think that almost anything which would confine the appetite within reasonable bounds would be regarded as a distinct blessing in this particular section. No hotel man can prosper

HOTEL SAGAMORE, LONG LAKE.

WEST SHORE RAILROAD,

New York Central & Hudson River R.R. Co. Lessee,

Superbly Built and Magnificently Equipped Double Track Steel Rail Trunk Line along the west shore of the world-famed HUDSON RIVER, and through the BEAUTIFUL MOHAWK VALLEY.

The popular route for business and pleasure travel between NEW YORK, ALBANY, BUFFALO AND NIAGARA FALLS. Forming, in connection with the *famous Hoosac Tunnel Line* and the *popular Great Western Division* of the Grand Trunk Railway, *the shortest route* between BOSTON, BUFFALO, NIAGARA FALLS, DETROIT, CHICAGO and ST. LOUIS.

MAGNIFICENT SLEEPING CARS are run regularly between Boston and Chicago ; Boston and Detroit ; Boston and St. Louis ; New York and Chicago ; New York, Buffalo, Niagara Falls and Detroit ; New York, Utica, Syracuse and Rochester.

PALACE DRAWING-ROOM CARS will be run during summer months between Boston and Syracuse ; New York and Buffalo ; New York and Albany ; New York and Saratoga ; New York and Caldwell (Lake George) ; New York, Phœnicia and Summit in Catskill Mountains ; Long Branch and Saratoga.

Direct All-Rail and only Drawing-Room Car Line to the Catskill Mountains.
Most Direct and Quickest Route to Lakes Mohonk and Minnewaska.

Trains on West Shore Railroad arrive at and depart from up-town station foot of West 42d St., and down-town station foot of Jay St., New York, and at Jersey City station of Penn. Railroad, making direct connection for Philadelphia, Baltimore, Washington, Long Branch and all New Jersey coast resorts, avoiding transfer through New York.

For information not obtainable at offices of West Shore Railroad or offices of connecting lines, address—

C. E. LAMBERT, *General Passenger Agent,*

5 Vanderbilt Avenue, New York.

surface. It is so like the pellucid run that drains the great Floridian Silver spring, that one almost expects to see some drowsy saurian roll off in the stream or glide away into the hammock, but instead we only scare away down the river a noisy picket guard of ducks. There is a "carry" at Raquette Falls and a rude hotel along here, known to the world as Mother Johnson's.

Presently there is a transformation. Away down the river the lumbermen a few years ago built a dam to hold this bright little river in check, the waters flowed back upon the forest and sapped its life. Fire sweeps away the giants of the wood, as it passes by, but it prepares the ground for a new growth. Water makes an eternal graveyard. The balance of the way to the Adirondack "carry" is through a spectral army of drowned hemlocks.

A bunch of little lakes appear upon the maps as Stony Ponds, but are sometimes called the *Spectacles*.

John Duckett has divided his time for a generation between planting and harvesting a well-cleared farm and running his hotel, the Hiawatha House, with a good share of attention to the deer by way of variety. You may thank John Duckett and his sons, excellent guides all of them, for the horns of the buck you carry back to the outer world as a trophy of your rifle, but you must thank Mrs. Duckett, and her comely daughter, who bring you trout and coffee, for the good cheer indoors. Mr. Duckett has a number of keen hounds, and if you really want to try your hand at the deer you are commended to his services.

It is a mile "carry" to the upper Saranac. Corey's place stands upon the hill at the end, a long log structure of two floors and abundant piazzas. Our particular tour does not include the Saranacs, but the way is straight and easy through to Paul Smith's, and thence out by rail to the line of of the Odgensburg & Lake Champlain Railroad at Moira.

The Saranacs are not yet invaded by the vicious pickerel (or, more properly, *pike*), and trout are therefore plenty.

Leaving Duckett's we follow the Raquette through the cemetery of drowned trees a dozen miles, and dine at the foot of Tupper Lake (Mt. Morris House). Uncle "Mart" Moody sold out last year and had got the lumber together half a mile above for a new house where his friends will probably find him, when they need either 'his services or his philosophy.

Tupper is as full of islands as a country belle's face with freckles. Near its head is the widely known Tupper Lake House, which is run by a former invalid city-hotel man, Mr. W. K. McClure. It is only second to the great Prospect House in the surprising completeness of its comforts, an oasis of luxury in the wilderness. Half a mile away Bog River plunges over a ledge, like a wilful suicide, straight into the lake. We put our Adirondacks into the stream just above and are away once more through glades unharmed by human touch. The chances are good of catching a glimpse of a deer, especially if you don't happen to have a gun.

It is curious how many strange and unexpected creatures a man *will* come upon when he has left his gun at home.

Along here you may begin to brace yourself for a "Two-mile carry" which intervenes before

BOG RIVER FALLS.

we may float upon the placid bosom of Little Tupper. While we are sorting out the several "traps" which are to be lugged over the snake-path ahead, our guide, while tying his oars, relates a dismal tale of a woodsman who blew his head off with a gun right here a few years back.

An hour later we come down to the gravelly shore of Round Pond, filled with the same devout thankfulness that is credited to De Soto and his copper-fastened army when they first looked upon the Mississippi.

An outlet leads to Little Tupper, also to a supper and bed at Pliny Robbins' Little Tupper House.

The still lake is furrowed and tossed into circling ripples by myriad speckled trout catching their food. It is too late to tempt them with a worm or fly ; but our guides have landed plenty o them for breakfast before we return from the realm of dreams.

No tourist who follows the round trip which these pages outline will have cause to complain of monotony or want of exercise during the day's voyage which follows his departure from Pliny Robbins, and which terminates at Forked Lake. It is as varied as a woman's whims. The longest portage is three miles and the shortest, that between Carry Pond and Little Forked Lake, the last of the series. The chain of ponds scattered like pearls along our course—Bottle, Sutton, Carry, etc.—are all deeply set in the forest, and full of fish. Bottle Pond, pictorially considered, is an artistic success, while its name, soothingly suggestive to him who hath a proper commissary, will always be attractive to the sportsman and wayfarer.

At Fletcher's, upon Forked Lake, we have completed the loop in our tour, and may look at our dishevelled, but sun-browned selves in the glass. The inspection ought to be a pleasing surprise.

If you are tired you won't be so in the morning. These jaunts beget unsuspected endurance, and the guides will tell you that hundreds of ladies pass over the same route every summer and enjoy it, too

WESTWARD, VIA THE FULTON CHAIN.

A short distance to the south of Raquette Lake is Eighth Lake, reached by the Brown Tract Inlet and a portage, which is the key to an outlet westward to Boonville upon the Eastern Division of the Rome, Watertown and Ogdensburg Railroad.

The way leads along Eighth, Seventh, Sixth, Fifth, Fourth, Third, Second and First Lakes.

These lakes are all celebrated for the plentitude of the fish, and are either connected by

A HALF HOUR'S TROUTING.

inlets or short portages. These are collectively known as the Fulton Chain. Wallace refers to this series in the following quotation:

Where, within the limits of the Adirondack region, can be found a brighter array of glittering links than the Fulton chain? Where a much lovelier sheet than Smith's Lake? Headley manifested his true appreciation of this section when he wrote the following:

"The Eight lakes are connected by streams, and form a group of surpassing beauty. They vary, both in size and shape, each with a different frame-work of hills, and the change is ever from beauty to beauty.

"There they repose like a bright chain in the forest, the links connected by silver bars. You row slowly through one to its outlet, and then entering a clear stream overhung with bushes, or fringed with lofty trees, seem to be suddenly absorbed by the wilderness. At length, however, you emerge as from a cavern, and lo! an untroubled lake, with all its variations of coasts, timber and islands, greets the eye.

"Through this you also pass like one in a dream, wondering why such beauty is wasted where the eye of man rarely beholds it."

The Forge House is a favorite half-way hotel upon this tour.

The water tour may be abandoned either here or at Arnold's (two and a half miles beyond by the road) in favor of the buckboard which is to convey you out through the "John Brown Tract" to civilization at Boonville, a further journey of twenty-five miles. The entire trip from the Raquette out is about fifty-two miles.

As a matter of fact, by far the larger portion of travelers over this route *come in* at Boonville.

An effort will be made in coming seasons to popularize this route betwen the Black River and headwaters of the Hudson, via Blue Mountain Lake, with tourists and summer residents returning from the Thousand Islands.

Indeed, there seems ample reason to expect a liberal response to the energy and enterprise which is now engaged in improving the trails and waterways that intervene.

This trip can be comfortably made in three days either way between Boonville and Saratoga, with good stopping places at night.

FROM NEW YORK TO THE MOUNTAINS.

It is due to the reader to suggest the several routes by which the connection with the Adirondack Railway at Saratoga may be reached. For special hotel information see pages following the time-tables.

Leaving early in the morning by rail over either the New York Central & Hudson River R. R. or the West Shore, connection is made at Albany with the north-bound train over the Delaware & Hudson Canal Co.'s Railroad, the same cars going through from the metropolis to Saratoga. A through sleeping car also leaves New York at 6.30 p. m. over the former route, arriving at North Creek early in the morning, and Blue Mountain Lake in time for early dinner.

Many travelers will prefer the pleasant evening voyage up the Hudson upon one of the uperb steamers of the People's Line to Albany or Citizen's Line to Troy, with a restful night and good connection with the train for Saratoga in the morning.

The Day Line steamers are also very popular with the traveling public, and trains connect with them as well as with the other lines for Saratoga.

TABLE OF STATIONS AND DISTANCES, ADIRONDACK R. R.

DISTANCE.	STATION.	DISTANCE.	STATION.	DISTANCE.	STATION
...	SARATOGA. DEL. & HUD. DEPOT.	13	SOUTH CORINTH	30	STONY CREEK
...	JUNCTION.	17	JESSUP'S LANDING	36	THURMAN
		...	SACANDAGA RIVER	44	THE GLEN
6	GREENFIELD.	22	HADLEY (LUZERNE)	50	RIVERSIDE
10	KING'S.	27½	QUARRY SWITCH	58	NORTH CREEK

For any further information apply at

Office of THE ADIRONDACK RAILWAY COMPANY, 20 Nassau Street, New York,

OR TO C. E. DURKEE, SUPERINTENDENT, SARATOGA SPRINGS, N. Y.

Send 4c stamp for additional copies of this book to either of the above.

ADIRONDACK RAILWAY COMPANY EXCURSIONS.

Saratoga to Jessup's Landing and Return (good for one day only),	$1 00
Saratoga to Hadley (Luzerne) and Return (good for one day only),	1 25
Saratoga to Schroon Lake and Return (good for four days only),	5 50
Saratoga to Blue Mountain Lake and Return (good for ten days only),	8 25
Saratoga to Raquette Lake or Forked Lake carry and Return (good for ten days only),	10 25
Hadley to Schroon Lake and Return (good for four days only),	4 25
Hadley to Blue Mountain Lake and Return (good for ten days only),	7 25
Hadley to Saratoga and Return (good for one day only),	1 25

LAKE LUZERNE EXCURSION.

ONE DAY EXCURSION TICKETS TO

HADLEY (STATION FOR LUZERNE) AND RETURN,

Are on Sale at all Ticket Offices in Saratoga, at the VERY LOW RATE OF **$1.25**.

Leave Saratoga.....	1.15 a.m.	10.10 a.m.	3.10 p.m.
Arrive Hadley (Station for Luzerne)...............	2.45 "	11.23 "	4.11 "

RETURNING

Leave Hadley...	7.23 a.m.	11.15 a.m.	4.33 p.m.
Arrive Saratoga...............	8.25 "	12.25 "	5.45 "

The GAME and FISH DINNERS furnished at the excellent hotels in Luzerne are renowned the world over. Beautiful drives about Luzerne, and river and mountain scenery unsurpassed.

BLUE MOUNTAIN AND RAQUETTE LAKE EXCURSION.

TICKETS GOOD FOR TEN DAYS.

Saratoga to Raquette Lake and Return, only $10.25.

A fine opportunity to tourists desirous ot visiting these beautiful lakes, so justly called the "GEMS OF THE ADIRONDACKS."

TIME TABLE AND DESCRIPTION OF ROUTE.

LEAVE SARATOGA at 10.10 a.m. via Adirondack Railway Co.; arrive North Creek 1.00 p.m.; leave North Creek (first-class four and six horse Concord Coaches and easy riding covered Buckboards) at 1.10 p.m., dining at North River (5 miles from North Creek) and arriving at Cedar River about 5 p.m., and at Blue Mountain Lake about 7 p.m. Next morning leave Blue Mountain Lake at 9 a.m., on steamer "Towahloonda," passing through Blue Mountain Lake, Eagle Lake, Utowana River and Utowana Lake, to Marion River carry, taking the steamboat "Killoquah" at the landing on Marion River for points on the "**Queen of American Lakes**," the beautiful Raquette. Steamboats leave passengers at Bennett's at 11.40 a.m., Kenwill's at 11.55 a.m., and arrive Forked Lake carry at 12.45 p.m., in season for early dinner at Forked Lake Hotel

RETURNING:

LEAVE FORKED LAKE CARRY at 2.40 p.m. by same route, arriving at Blue Mountain Lake at 6.35 p.m. Next day leave Blue Mountain Lake at 7 a.m., and arrive at North Creek at 2.50 p.m.; leave North Creek at 3 00 p.m., arriving at Saratoga at 5.45 p.m. The steamboats "Towahloonda" and "Killoquah" were built expressly for carrying passengers on these lakes, and are roomy, comfortable and perfectly safe. The "Killoquah" will land and pick up passengers on Raquette Lake at Bennett s, Kenwill's and Forked Lake carry.

SCHROON LAKE EXCURSION TICKETS

To Schroon Lake and return, good for four days, can be purchased at all ticket offices in Saratoga at the low rate of $5.50.

Holders of tickets can stop over, if they choose, at LAKE LUZERNE, and at any other point on the route, provided they use their tickets through to destination within FOUR DAYS from date of sale. Visitors should not leave Saratoga without making a trip to these beautiful lakes.

TIME-TABLE BETWEEN SARATOGA AND SCHROON LAKE.

	SLEEPING CAR.					
Leave Saratoga (Adirondack R'y).........	1.15 a.m.	10.10 a.m.	3.14 p.m.	Leave Pottersville (steamboat)	3.00 p.m.	7.30 p m.
Arrive Riverside....	4.35 "	12.40 p.m.	5.50 "	Arrive Adirondack....	3.25 "	7.40 "
Leave " (stages)........	....	12.55 "	6.00 "	" Lake View Point	3.40 "	7.50 "
Arrive Pottersville (dine)	....	2.10 "	7.10 "	" Schroon Lake Village..........	4.15 "	8.35 "

TABLE OF DISTANCES.

	MILES		MILES
Saratoga to Hadley (Luzerne)............................	22	North Creek to Lewey Lake.............................	22
" Riverside	50	" Chain Lakes......	22'
" North Creek......	58	" Blue Mountain Lake	29
Luzerne to Lake George.	10	Blue Mountain Lake to Eagle Lake...	2½
Riverside to Chestertown....	6	" " " Utowana Lake.........	3½
" Pottersville	7	" " " Marion River Carry.........	7
Pottersville to Lake View..	4	" " " Raquette Lake, Bennett's...	13
" Adirondack....	4	" " " " Kenwills......	15
" Schroon Lake Village	9	" " " " Hathorns... ...	15
North Creek to Olmsteadville	8½	" " " Forked Lake Carry......	21
" Aiden Lair..	16½	" " " Forked Lake......	21½
" Newcomb........	22	" " " Little Forked Lake..............	23½
" Tahawus	27	" " " Brown's Tract Pond...	17
" Adirondack Iron Works...	37	" " " Shed Lake........	17
" Long Lake...	44	" " " Eighth Lake, Fulton Chain............	17
" North River...	4	" " " Long Lake Village, via Raquette and	
" Thirteenth Pond...	8	Forked Lake......	31
" Indian River	15	" " " Long Lake Village, via Highway	10
" Indian Lake...	16	" " " Brandreth Lake...	19
" Cedar River.....	20		

TABLE OF DISTANCES.

Excursions Blue Mountain Lake to Saranac Lake and to Tupper's Lake and Return.

	Distance from Blue Mountain Lake.	
	MILES.	
Blue Mountain Lake, through Eagle and Utowana Lake to Marion River carry	6	6
Over carry	⅓	6⅓
Through Raquette Lake to Forked Lake carry	12	18⅓
Over carry	½	18¾
Through Forked Lake to carry on Raquette River	4	22¾
Over carry	1½	24¼
Raquette River to Butler, with Falls carry	1½	25¾
Carry around Falls	⅛	
Raquette River to carry	½	26½
Over carry to Long Lake	½	27
Through Long Lake to Kellogg's	4	31
Kellogg's, through Long Lake to Mother Johnson's carry	17	48
Around carry	1½	49½
Through Raquette River to Indian carry (Duckett's)	10	59½
Across Indian carry to Saranac Lake (Cory's)	1	60½
Via Lake to Bartlett's	2	62½
Bartlett's, via Lake to Martin's	12	74½
Bartlett's, via Lake to Prospect House	8	70½

	Distance from Blue Mountain Lake.	
	MILES.	
Prospect House (Saranac Lake) to Bloomingdale	19	89½
" " to Paul Smith's by water	10½	81
Bloomingdale " " road	10	90½
Paul Smith's to Malone on O. & L. C. R.R.	57	136
Blue Mtn. Lake to Mother Johnson's carry		48
From Mother Johnson's carry, through Raquette River to Moody's, Big Tupper Lake	18	66
Through Big Tupper Lake to Tupper Lake House	7	73
" streams and carrys to Round Pond	4¾	77¾
" Round Pond	2½	80¼
Over carry to Little Tupper Lake	1	81¼
Through Little Tupper Lake	6	87¼
Over carry to Rock Pond	2	89¼
" " Bottle Pond	3	92¼
To Cary Pond	2	94¼
Carry to Little Forked Lake	1	95¼
Little Forked Lake to Steamboat Landing	6	101¼
" " " Blue Mountain Lake	18¾	120

TABLE OF ELEVATIONS.

	FEET ABOVE TIDE WATER.			FEET ABOVE TIDE WATER.
Depot Saratoga Springs	306	Upper Saranac Lake		1,567
Hadley (Luzerne)	615	Lower " "		1,527
Riverside	875	Tupper's Lake		1,500
North Creek	990	Indian Lake		1,705
North River	1,041	Long " "		1,614
Rock River Dam	1,760	Saranac Lake		1,539
Blue Mountain Lake	1,800	Schroon Lake		830
Raquette Lake	1,774	Thirteenth Pond		1,953
Forked Lake	1,753	Tupper Lake, Big		1,554
Eighth Lake	1,803	" " Little		1,728
Beach's Lake	1,861			

─FARES─

(SUBJECT TO CHANGE.)

New York to Hadley (Luzerne) via N. Y. C. & H. R. R. R., or West Shore R.R.......... $5.15
    "   "   "   "   " Day Line Steamers........ 4.20
    "   "   "   "   " People's Line Steamers.... 3.70
    "   "   "   "   " Citizens' "   " 3.50
    " " Schroon Lake " N. Y. C. & H R. R. R., or West Shore R. R..... 7.90
    "   "   "   "   " Day Line Steamers 6.95
    "   "   "   "   " People's Line Steamers... 6.45
    "   "   "   "   " Citizens' "   "   ... 6.25
    " " Blue Mtn. Lake " N. Y. C. & H. R. R. R., or West Shore R. R. 9.45

New York to Blue Mtn. Lake via Day Line Steamers........ $8.45
  "   "   "   "   "   " People's Line Steamers.... 7.95
  "   "   "   "   "   " Citizens' "   " 7.75
  "   " Raquette Lake " N. Y. C. & H. R. R. R., or West Shore R. R........ 10.95
  "   "   "   "   " Day Line Steamers........ 9.95
  "   "   "   "   " People's Line Steamers .. 9.70
  "   "   "   "   " Citizens' "   " ... 9.25
Boston to Hadley (Luzerne)............................. 6.15
  " " Schroon Lake..... 8.90
  " " Blue Mountain Lake............................. 10.40
  " " Raquette Lake............................. 11.65

Via Pennsylvania or Philadelphia & Reading R. R. from **PHILADELPHIA AND POINTS SOUTH** to New York, N. Y. C. & H. R. R. R. ; or West Shore R. R. to Troy ; or Albany, D. & H. C. Co. to Saratoga, Adirondack Railway, Stages and Steamboats to destination.

FROM	TO LUZERNE.	TO SCHROON LAKE.	TO BLUE MTN. LAKE.	TO RAQUETTE LAKE.
Philadelphia, Pa	$7.65	$10.40	$11.90	$13 40
Baltimore, Md....	10.45	13.20	14.70	16.20
Washington, D.C	11.65	14.40	15.90	17.40

Via Day Line Steamers New York to Albany.

FROM	TO LUZERNE.	TO SCHROON LAKE.	TO BLUE MTN. LAKE.	TO RAQUETTE LAKE.
Philadelphia, Pa........	$6.70	$9.45	$10.95	$12.45
Baltimore, Md...	9.50	12.25	13.75	15.25
Washington, D.C...........	10.70	13.45	14.9	16.45

Via People's Line Steamers New York to Albany.

FROM	TO LUZERNE.	TO SCHROON LAKE.	TO BLUE MTN. LAKE.	TO RAQUETTE LAKE.
Philadelphia, Pa...	$6.20	$8.95	$10.45	$11.95
Baltimore, Md..............	9.00	11.75	13.25	14.75
Washington, D.C..	10.20	12.95	14.45	15.95

Via Citizen's Line Steamers New York to Troy,

FROM	TO LUZERNE.	TO SCHROON LAKE.	TO BLUE MTN. LAKE.	TO RAQUETTE LAKE.
Philadelphia, Pa	$6.00	$8.75	$10.25	$11.75
Baltimore, Md...	8.80	11.55	13.05	14.55
Washington, D.C	10.00	12.75	14.25	15.75

around here who depends upon a profit from his table, while the promiscuous distribution of a couple of pounds of the commodity means swift and certain ruin. Its popularity is said to be great among the young people, who relieve the embarrassment of an awkward *hiatus* in conversation by its constant mastication.

Long Lake is more a wide and peaceful river than a lake. It is in every sense of the word a beautiful sheet of water, and the summer retreats of the Rev. Dr. Duryea, of Boston, and those of Senator Platt, and others, perched upon its shores, speak volumes for its attractiveness.

Long Lake extends its attenuated form thirteen miles, and then we come to another picturesque link in the Raquette River, here reinforced by the waters of Cold river draining the slope of Mt. Seward. Mile after mile we go, the swift current giving us double impetus, and supper at Duckett's seems assured This is a surpassingly lovely water-lane. Each over-reaching bough, gaunt and withered cedar spine, and trailing vine is mirrored in the still but shifting

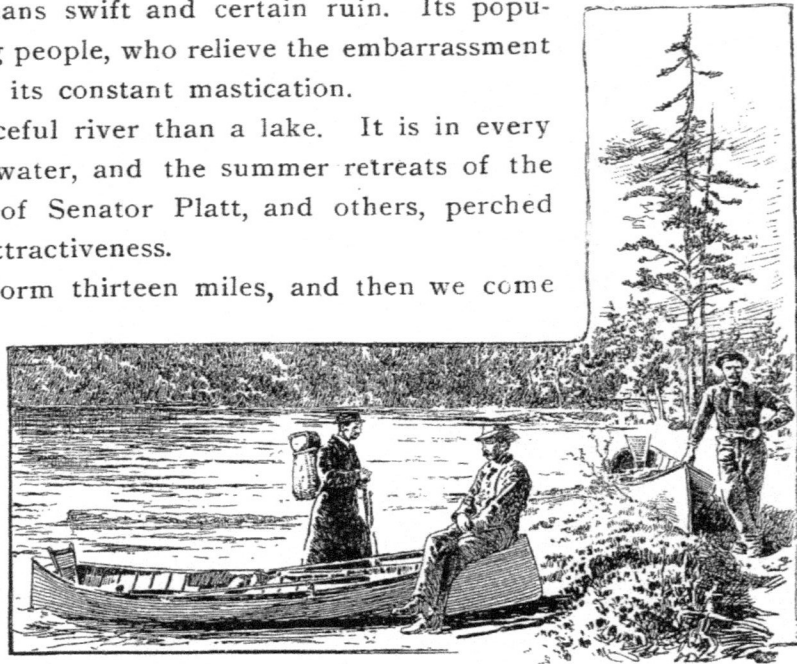

ALONG THE RAQUETTE.

EXCURSION RATES FROM NEW YORK.

(SUBJECT TO CHANGE.)

To Lake Luzerne, N. Y.

| Via N. Y. C. & H. R. R. R. | - | - | - | to Troy. | Del. & Hudson Canal Co. R. R. | - | - | to Saratoga. |
| or West Shore R. R. | - | - | - | " Albany. | Adirondack Railway Co. | - | - | - | " Hadley. |

Returning by same route, $9.50.

To Riverside, N. Y.

| Via N. Y. C. & H. R. R. R. | - | - | - | to Troy. | D. & H. C. Co's. R. R. - | - | - | - | to Saratoga. |
| or West Shore R. R. | - | - | - | " Albany. | Adirondack Railway | - | - | - | - | " Riverside. |

Returning by same route, $11.50.

To Schroon Lake, N. Y.

Via N. Y. C. & H. R. R. R.	-	-	-	to Troy.	Adirondack Railway	-	-	-	-	to Riverside.
or West Shore R. R.	-	-	-	" Albany.	Stages	-	-	-	-	" Pottersville.
D. & H. C. Co's. R. R.	-	-	-	" Saratoga.	Steamers	-	-	-	-	" Schroon Lake.

Returning by same route, $15.00.

To Blue Mountain Lake, N. Y.

Via N. Y. C. & H. R. R. R.	-	-	-	to Troy.	Adirondack Railway	-	-	-	-	to North Creek.
or West Shore R. R.	-	-	-	" Albany.	Blue Mtn. Lake Stage & Transportation Co. (stages)	" Blue Mtn. Lake.
D. & H. C. Co's. R. R.	-	-	-	" Saratoga.		

Returning by same route, $16.25.

To Forked Lake Carry, N. Y.

Via N. Y. C. & H. R. R. R.	-	-	-	to Troy.	Blue Mtn. Lake Stage & Transportation Co.	-	to Blue Mtn. Lake.
D. & H. C. Co's. R. R.	-	-	-	" Saratoga.	Blue Mtn. Lake Steamboat Co.	-	-	" Forked Lake
Adirondack Railway.				" North Creek.			Carry

Returning by same route, $18.75.

SUMMER EXCURSION RATES.

(SUBJECT TO CHANGE.)

To Luzerne, N. Y.

Via N. Y. C. & H. R. R. R., or West Shore R. R., in both directions.

Philadelphia & Reading or Pennsylvania R. R. - to New York.	Delaware & Hudson R. R. - - - - to Saratoga.
N. Y. C. & H. R. R. R. - - - - - " Troy.	Adirondack Railway " Hadley, (Luz.)
West Shore R. R. - - - - - - - " Albany.	

Returning by same route.

THROUGH RATES.

Baltimore, Md.	$19.10	Richmond, Va.	$28.50
Philadelphia, Pa.	13.50	Washington, D. C.	21.50

To Schroon Lake, N. Y.

Via same route to Saratoga and Adirondack Railway to Riverside.

Stages, 7 miles to Pottersville. - - - - Steamers, 9 miles to Schroon Lake.

Returning by same route.

THROUGH RATES.

Baltimore, Md.	$24.60	Richmond, Va.	$34.00
Philadelphia, Pa.	19 00	Washington, D. C.	27.00

To Blue Mountain Lake, N. Y.

Via N. Y. C. & H. R. R. R., or West Shore R. R. in both directions; or Philadelphia & Reading R. R.

Pennsylvania R. R. '- - - - to New York.	Delaware & Hudson R. R. - - to Saratoga.
N. Y. C. & H. R. R. R. - - - - - " Troy.	Adirondack Railway - - - - " North Creek.
West Shore R. R. - - - - - - " Albany.	Stages, 30 miles " Blue Mtn. Lake

Returning by same route.

THROUGH RATES.

Baltimore, Md.	$25.85	Richmond, Va.	$35.25
Philadelphia, Pa.	20.25	Washington, D. C.	28.25

Via New York and People's Line Steamers in both directions, returning by same route,	$2.60 less than above rates.
" Citizen's Line Steamers,	3.00 " "
" Day Line Steamers,	1.60 " "

CONDENSED TIME TABLE.
(SUBJECT TO CHANGE.)

NORTH BOUND.

Lve. Washington	11.15 a.m.	4.15 p.m.	11.20 p.m.
" Baltimore	12.18 p.m.	5.15 "	12.35 a.m.
" Philadelphia	3.00 "	7.49 "	3.45 "
Lve. Boston	3 00 p.m.	7.00 p.m	
" " via B.& A. or Fitch.R.R		10.30 "	
Lve. New York	6.30 p.m.	11.15 p.m.	9.00 a.m.
" Peekskill	7.48 "	12.55 "	
" Garrisons	8.06 "		10.19 "
" Poughkeepsie	9.05 "	2.38 a.m.	11.05 "
" Albany	11 40 "	8.25 "	1.15 p.m.
" Troy	11.50 "	8.25 "	1.25 "
Arr. Saratoga	1.05 a.m.	9.50 "	2.45 "
Lve. Schenectady		6.50 a.m.	1 50 p.m.
Arr. Saratoga		7.35 "	2.40 "
Lve. Saratoga	1.15 a.m.	10.10 a.m.	3.10 p.m.
" Jessups Landing	2.20 "	11.00 "	3.58 "
" Hadley (Luzerne)	2.40 "	11.19 "	4.11 "
" Riverside	4.20 "	12.37 "	5.30 "
" North Creek	4.45 "	1.00 "	5.52 "
" Schroon Lake	9 25 "	4.15 "	8.40 "
" Blue Mountain Lake	1.15 p.m.	7.15 "	

SOUTH BOUND

Lve. Blue Mtn. Lake				7 30 a.m.
" Schroon Lake			6.45 a.m.	11.45 "
" North Creek	6.00 a.m.			3.00 p.m.
" Riverside	6.20 "		9.55 "	3.25 "
" Hadley (Luzerne)	7.30 "		11.15 "	4.33 "
" Jessups Landing	7.43 "		11.31 "	5.06 "
Arr. Saratoga	8.30 "		12.20 p.m.	5.45 "
Lve. Saratoga	8 40 a.m.	12.50 p.m.	6.15 p.m.	6 25 p.m
Arr. Troy	9.40 "	2.05 "	7.28 "	
" Albany	9.50 "	2.15 "		7.50 "
Lve. Troy via H. R. R. R	9.45 a.m.	2.25 p.m.	2.30 a.m.	
" Albany	9.50 "	2.40 "	3.10 "	
" Poughkeepsie via H. R. R. R	12.00 m.	4 50 "	5.20 "	
Arr. New York	2.12 p.m.	7.00 "	8.00 "	
Lve. Saratoga	8.55 a.m.	12 35 p m	10.00 p.m.	
Arr. Schenectady	9.40 "	1.20 "	10.45 "	
" New York via West Shore R. R.	3.05 p.m.	7.25 "		
Lve. New York via Penn. R. R	3.40 p m	9 00 p.m	10 00 a.m.	
Arr. Philadelphia	5.55 "	12.10 "	12.25 p.m.	
" Baltimore	8.55 "	3.50 "	2.57 "	
" Washington	10.10 "	5.35 "	4.00 "	

Connections from New York by Night Boats on Hudson River.

PEOPLE'S LINE.—From Pier 41, North River, daily, except Sunday, at 6.00 p.m., arriving at Albany at 6.00 a.m. Immediate fast steamboat train to SARATOGA and LAKE GEORGE.

CITIZEN'S LINE.—From Pier 44, North River, daily, except Saturday, at 6.00 p.m., arriving at TROY at 6.30 a.m.

——o——

CONNECTIONS FROM NEW YORK.

THE HUDSON RIVER BY DAYLIGHT.
Steamers "ALBANY" and "CHAUNCEY VIBBARD."

Leave New York daily, except Sunday, as follows:

From Pier 39, North River, foot of Vestry Street.... ...	8 40 a.m.
From Pier foot of 22d St., North River.	9 00 "
Arrive Albany......	6 10 p.m.

SPECIAL NOTICE.—The time-table of the New Stage Line from North Creek to Blue Mountain Lake, is so arranged that passengers arrive at Blue Mountain Lake before dark. Passengers taking special buckboards from North Creek for Blue Mountain Lake can remain in sleeper till 9 a.m. For special buckboards apply to Supt. Stage Line, North Creek, N. Y. Passengers for Schroon Lake may remain in sleeping car and go through to North Creek and return on 6 a.m. train to Riverside. Baggage checked through to Blue Mountain Lake from New York and Philadelphia; also to Schroon Lake, Luzerne, and other points on line of Adirondack Railway Co.

BOSTON via ALBANY.

Lve. Saratoga	8.40 a.m.	12.50 p.m.	6.25 p.m.
" Albany	10.00 "	2.37 "	8.40 "
Arr. Springfield	1.35 p.m.	6.35 "	12 40 a.m.
" Boston	4.35 "	9.45 "	6.25 "

BOSTON via RUTLAND.

Lve. Saratoga 10.00 a.m. . . . . Arr. Boston 6.35 p.m.

Time-Table.—The Adirondack Railway Company.

SEASON OF 1887. Subject to changes that may be made in D. & H. C. Co.'s Time-Table.

TRAINS GOING NORTH.				TRAINS GOING SOUTH.			
STATIONS.	1 Mail and Express.	3 EXPRESS.	7 Blue Mountain Lake Special.	STATIONS.	2 New York Special.	4 EXPRESS.	6 Mail and Express.
Saratoga	10.10 a.m.	3.14 p.m.	1.15 a.m.	North Creek	6.00 a.m.	9.40 a.m.	3.00 p.m.
Greenfield	10 33 "	3.38 "	1.45 "	Riverside	6 17 "	10.00 "	3.22 "
Kings	10.43 "	3.48 "	1 58 "	The Glen	6.30 "	10.16 "	3.38 "
South Corinth	10 53 "	3.58 "	2.10 "	Thurman	6.48 "	10.36 "	4.00 "
Jessup's Landing	11.03 "	4.09 "	2.25 "	Stony Creek	7.02 "	10 52 "	4.15 "
Hadley (Luzerne)	11.23 "	4.35 "	2 45 "	Hadley (Luzerne)	7.23 "	11.15 "	4.39 "
Stony Creek	11 46 "	4.58 "	3 15 "	Jessup's Landing	7.37 "	11.34 "	4.55 "
Thurman	12.03 p.m.	5.13 "	3 37 "	South Corinth	7.45 "	11.44 "	5.05 "
The Glen	12.23 "	5.33 "	4.10 "	Kings	7 53 "	11.53 "	5.14 "
Riverside	12.40 "	5.50 "	4.35 "	Greenfield	8.03 "	12.03 p.m.	5 24 "
North Creek	1.00 "	6.10 "	5.00 "	Saratoga	8.25 "	12.25 "	5.45 "

Passengers arriving North Creek in sleeping car at 5 a.m. can remain in car till 9 a.m.   All trains start from and stop at Delaware & Hudson Canal Co.'s Depot at Saratoga.   Trains 7, 9, 8 and 10 will run only during July and August.

STAGE CONNECTIONS.—At Jessup's Landing for Corinth (Palmer's Fall's) ; at Hadley for Conklingville and the Sacandaga Valley ; at Stoney Creek for Creek Centre ; at Thurman for Warrensburgh ; at Riverside for Weavertown, Johnsburgh, Chester, Pottersville and Schroon Lake ; at North Creek for North River. Indian River, Cedar River. Blue Mountain Lake and Long Lake.

LUZERNE, SCHROON AND BLUE MOUNTAIN LAKES.

THROUGH CAR SERVICE BETWEEN

NEW YORK and NORTH CREEK, via H. R. R.R.

A Wagner Sleeping Car will leave New York 6.30 p.m. daily (except Sunday), running through to North Creek without change, arriving at North Creek about 5 o'clock next morning.   Passengers will be allowed to remain in cars till 9 a.m.

LOCAL DRAWING ROOM CAR SERVICE BETWEEN SARATOGA AND NORTH CREEK.

Wagner Drawing Room Car on train leaving Saratoga 10.10 a.m.   Returning, leave North Creek 3.15 p.m.

The Adirondack Railway Company.

SPECIAL NOTICE.

New Arrangement of South Bound Trains.

The New York Special will leave North Creek at **7 p. m., daily,** on arrival of stages from Blue Mountain Lake, instead of **6 a. m.,** as published in Time Table. A through sleeping Car will be attached to this train at North Creek, arriving New York 7 a. m. next morning. Passengers can leave Raquette Lake in the morning, and Blue Mountain Lake after lunch, and connect with this train.

C. E. DURKEE, Superintendent.

Blue Mountain and Raquette Lake Steamboat Line.

Steamers "KILLOQUAH," "UTOWANA," "TOWAHLOONDA" and "IROCOSIA,"

WILL RUN AS FOLLOWS, UNTIL FURTHER NOTICE.

IN EFFECT JULY 1ST, 1887.

STEAMERS ON BLUE MOUNTAIN LAKE.			STEAMERS ON RAQUETTE LAKE.		
Lve. Holland's Hotel........	8.45 a.m.	2.45 p.m.	Lve. Forked Lake Carry.	8.30 a.m.	2.40 p m.
" Prospect House....	9.00 "	3.00 "	Arr. Kenwill's....................... ...	9.15 "	3.25 "
Arr. Marion River Carry........	10.00 "	4 00 "	Lve. Kenwill's........	9.20 "	3 30 "
STEAMERS.			Arr. Bennett's Dock, ¶.......................	9 30 "	3.40 "
Lve. Marion River Carry	10.40 "	4.55 "	Lve. Bennett's Dock...	10.35 "	3.45 "
Arr. Bennett's Dock, ¶...................	11.40 "	5.55 "	Arr. Marion River Carry.............	10.35 "	4.45 "
Lve. Bennett's Dock.........................	11.45 "	6.00 "	STEAMERS.		
Arr. Kenwill's............................	11.55 "	6.10 "	Lve. Marion River Carry	11.15 "	5.20 "
Lve. Kenwill's...........	12.00 m.	6.15 "	Arr. Blue Mountain Lake, Prospect House..	12.15 "	6.20 "
Arr. Forked Lake Carry.....................	12.45 p. m.	7.00 "	" " " " Holland's Hotel..	12.30 "	6.35 "

¶ Passengers for HATHORN's and south part of RAQUETTE LAKE, land at Bennett's Dock.

☞ Tickets can be had at Steamboat Office, Blue Mountain Lake.

HENRY BRADLEY, Superintendent,

BLUE MOUNTAIN LAKE, HAMILTON COUNTY, N. Y.

FROM BOSTON AND EASTERN POINTS.

The direct route from Boston, Providence and Worcester to Saratoga, and only route connecting with Adirondack R.R. in same station at Saratoga, is the old and well-known BOSTON & ALBANY R.R.

Drawing Room Cars from Boston to Saratoga during the season.

IMPORTANT TO TOURISTS.

TICKETS for Blue Mountain Lake may be procured at principal ticket offices in New York and Philadelphia, and baggage will be checked through. Passengers in Sleeping Car, arriving at North Creek at 5.00 a.m., may remain in car until 9 a.m.

Special Covered Spring Buckboards, North Creek to Blue Mountain Lake, seating three or four persons, can be had by telegraphing the SUPERINTENDENT, Blue Mountain Lake Stage and Transportation Company, North Creek, N. Y.

ADIRONDACK LAKES and MOUNTAINS

——VIA——

BLUE MOUNTAIN LAKE STAGE AND TRANSPORTATION CO.

SEASON OF 1887.

This entirely New Line, organized to meet the demands as to safety, speed and comfort of travelers between the North Creek terminus of the Adirondack Railway and Blue Mountain Lake and Long Lake, desires to call the Special attention of visitors to the mountains to the following facilities now offered to the public:

☞ *Through Tickets between New York and principal cities and Blue Mountain Lake are sold via Adirondack Railway over and* **ONLY** *over this Stage Line, and baggage is now checked through from the large cities to the Blue Mountain Lake direct and only over the same. No delay, no re-checking. Special covered Spring Buckboards, seating three or five persons, at a small additional charge, can be had by applying to the Superintendent of the Company at North Creek, Warren County, N. Y.*

Clean Stages and Buckboards, civil drivers and speedy conveyance of passengers, prompt delivery of baggage, and traveling done by daylight.

NEW ARRANGEMENT FOR 1887.

Blue Mountain Lake Stage and Transportation Company,

North Creek, Warren County, N. Y.

HOTELS AND BOARDING HOUSES.

JESSUP'S LANDING.—*Corinth, Saratoga Co., N. Y., 17 miles from Saratoga Springs.*—This village of about 600 inhabitants is on the west bank of the Hudson, about a mile east of the railroad station. It contains the extensive manufactory, Hudson River Pulp and Paper Company, for making pulp from wood for the manufacture of paper.

At Jessup's Landing is a magnificent waterfall in the Hudson, which here plunges over perpendicular rocks 70 feet in height. The falls, with the rapids extending half a mile above, afford a scene of remarkable grandeur and sublimity. At this point the traveller reaches the edge of the wild and mountainous Adirondack region.

CORINTH, POST OFFICE CORINTH, N. Y. (Jessup's Landing Station.)

S. P. Flagler, Hotel ; 1 mile Jessup's Landing Station ; conveyance stage, or private if desired ; accommodate, 50 : sleeping rooms, 24 ; rates, adults, $8.00 to $12.00 ; children, $4.00 to $6.00 ; transient, $2.00 per day ; discount to season guests. 3 stories high ; back and front piazza ; pleasant rooms ; newly fitted up ; first-class livery ; boating and fishing, Hudson River, Effnor's Lake and Hunt's Lake.

HADLEY, P. O. HADLEY, N. Y. (Hadley Station.)

John Halloran, Proprietor ; Railroad House ; accommodates 40 ; rates, $3.00 to $12.00 per week, $2.00 per day.

LUZERNE.—*Luzerne, Warren Co., N. Y., is on the east bank of the Hudson, a half mile from Hadley Station on the Adirondack Railway.*—This delightful little village is charmingly situated amidst some of the finest scenery to be found in Northern New York. The mountains on either side rise about 600 feet, and the boisterous Hudson plunges through a deep and rocky gorge, forming the most beautiful rapids and delightful prospects. 4 churches : Episcopal, Presbyterian, Methodist and Catholic.

LUZERNE, P. O. LUZERNE, WARREN CO., N. Y. (Hadley Station.)

H. J. Rockwell, Manager ; Wayside Hotel and cottages situated on an elevation above the village, and overlooking the lake ; extensive grounds ; shady walks ; one mile from station ; bus to all trains ; accommodation for 200 guests ; terms, $15.00 to $21.00 per week ; telegraph office in the house.

G. T Rockwell & Son, Rockwell's Hotel ; ½ mile (Hadley Station); transportation free ; bus to all trains ; accommodate, 150 ; rates, adults, $10.50 to $14.00 ; children, $7.00 to $10.50 ; transient, $3.00 per day ; abundance of splendid drives and walks ; good hunting and fishing ; milk, butter, and vegetables fresh daily from Rockwell Hotel Valley Farm ; convenient to post and telegraph offices ; references on application.

E. E Riddell, Hotel ; free carriage ; accommodate 100 ; sleeping rooms, 45 ; terms, transient, $2.25 ; adults, $10.50 to $14.00 ; children, $5.00 to $7.00 ; discount to season guests ; pleasant grounds ; extensive lawn for tennis and croquet, centrally located ; references, Dr. J.C. Hutchinson, 130 Hicks Street, Brooklyn ; P. H. Williams, 442 Madison Avenue, N. Y. City.

H. McMaster, Private Boarding, centrally located in village, about ¾ mile from station ; large and airy rooms ; cool and broad piazzas ; accommodation for 25 ; rates, $8.00 to $12.00 per week.

Mrs. D. A. McEwen, Private Boarding House ; ½ mile from station ; free carriage ; accommodate 20 ; 12 rooms ; adults, $8.00 to $10.00 per week ; children half price ; transient, $1.50 per day ; house centrally and finely located ; shady grounds ; easy of access to all points of interest ; references, 17 East 14th Street, N. Y. City ; 420 Gates Avenue, Brooklyn.

Mrs. M. O. Butler. Private Boarding ; ½ mile from station ; free carriage ; 20 rooms ; accommodates 20 ; rates, adults, $8.00 to $10.00 per week ; children, according to age ; discount for season June to September.

Miss M. J. Fisher, Private Boarding House ; ¾ mile from station ; accommodate 25 ; rates, $7.00 to $8.00 per week ; children, $3.00 to $6.00 ; large pine grove fronting house ; about 5 minutes' walk to the lake or river.

H. Beach, Farm House ; 2½ miles from station and village, on road to Lake George ; can accommodate 10 or 15 ; rates, $8.00 to $10.00 ; discount for the season. This cottage is pleasantly situated near cool and inviting pine groves, and is supplied with pure soft cold spring water.

CONKLINGVILLE, P. O. CONKLINGVILLE, SARATOGA CO., N. Y.

Thomas Salmon, Proprietor ; Cottage Hotel ; located in the Sacandaga Valley, 6 miles from Hadley station ; house situated about 200 feet from the bank of the river, and 700 feet above tide water ; can be reached by stage or private conveyance. Stages run on Tuesdays, Thursdays and Saturdays, fare 75 cents. House has accommodation for 30 ; rates, $7.00 per week ; transient, $1.50 per day. Good fishing in river and lakes near by ; kind of fish : trout, bass, pickerel ; also good gunning, game, patridge, woodcock and duck.

WARRENSBURGH, N. Y., P. O. WARRENSBURGH, WARREN CO., N. Y. (Thurman Station.)

C. H. Dickinson, Private House ; Three miles from Thurman Station ; daily stage ; accommodate six ; rooms 4 ; rates, $8.00 to $10.00 ; children, $3.00 to $5.00 per week.

Mrs. M. R. King, Farm House ; 3½ miles from Thurman Station ; pleasantly situated on a plain several feet above and overlooking the village, surrounded by pretty groves ; commands a view of both the Hudson and Schroon rivers ; daily stage from all trains ; accommodate 20 ; 12 sleeping rooms ; terms, $8.00 per week ; children, $3.00, upward ; discount for season.

John Heffron, Hotel ; accommodate 50 ; rates, $14.00 per week ; children, $7.00 ; transient, $2.00 per day.

CHESTERTOWN, N. Y., POST OFFICE CHESTERTOWN, N. Y.—Chestertown is a thriving little village of about 600 inhabitants, containing four churches—Methodist, Baptist, Presbyterian and Catholic. It has quite an elevation above the sea, and the surrounding country is picturesque, with its many little lakes, valleys, and rolling hills that at places rise into quite respectable mountains. It is six miles south of Pottersville ; the same east of Riverside, and eighteen north of Lake George.

G. W. Ferris, Chester House, Hotel ; 6 miles from Riverside Station ; 1,500 feet above tide ; reached by daily stage ; accommodation for 125 ; rates, for one week or less, $3.00 per day ; over one week, $2.00 per day ; children under 12 years, $7.00 per week ; liberal discount to families for the season. There is a fine grove near the house, also fine lawn tennis and croquet grounds. Table supplied with milk and vegetables fresh from the Hotel farm.

F. W. Rising, Rising House, Hotel ; accommodates 100 guests ; terms, $2.00 per day : $10.00 per week.

Royal P. Mead, Farm House ; 7 miles from Riverside Station ; accommodate 15 ; rooms 9 ; adults, $6.00 ; children under 10 years, $4.00 ; transient, $1.00 ; transportation by stage.

HOTELS AND BOARDING HOUSES—*Continued.*

SCHROON LAKE, POST OFFICE SCHROON LAKE, ESSEX CO., N. Y.

L. R. & E. D. Loche, Leland House ; has first-class accommodations for 200 guests ; has grounds 5 acres in extent, tastefully laid out ; commands a view of the lake, and the beautiful scenery of Schroon Valley. For terms, etc., see advertisement on another page.

J. M. Leland, Leland Cottage ; is pleasantly located on high dry ground, west side of village in Schroon Lake ; large yard in front ; both house and yard are beautifully shaded with maple trees ; can accommodate from 20 to 25 boarders ; price, from $6.00 to $9.00 per week ; for particulars address, J. M. Leland, Schroon Lake, Essex Co., N. Y.—J. M. Leland, drug and variety store on South Avenue, near Leland House. Doctor's office over drug store.

C. F. Taylor, Taylor House and cottages, twelve in number ; capacity 150 ; located at Lake View Point, in a pine grove, 4 miles from outlet of Schroon Lake ; commands a view of almost entire lake north and south. The Lake View farm supplies the usual vegetables of the season. The steamer touches at the dock on all regular trips. For rates, etc., see advertisement on another page. P. O. South Schroon, N. Y.

W. A. McKenzie, Jr., Grove Point House ; pleasantly situated in a shady grove on the west shore of the lake ½ a mile from the village ; rooms large and well furnished ; terms, $8.00 to $12.00 per week, $2.00 per day.

Windsor Hotel. The Windsor is a first-class summer hotel ; broad piazzas extend the entire length of the house ; beautiful shaded lawn for lawn tennis and croquet ; large airy rooms with commanding view of mountains and lake ; accommodate 100 ; rates, $2.00 to $2.50 per day $10.00 to $15.00 per week.

C. C. Whitney, the Arlington Cottage ; ½ mile north steamboat landing ; extensive views of lake and mountain scenery ; accommodate 15 ; $8.00 $12.00 per week.

W. S. Fowler, Fowler House ; accommodations for 50 ; board $7.00 to $12.00.

C. W. Burwell, Ondawa House ; capacity 100 ; rates, $10.00 to $12.00 per week, $2.00 per day.

Martin Smith, Paradox House, on south side of Paradox Lake ; 6 miles from Schroon Lake ; will accommodate 20 ; board $12.00 per week ; $2.00 per day ; specialty game and fish dinners at all hours.

Orrin Harris, Pyramid Lake House. Post Office Paradox, Essex Co., N. Y. Accommodation for 30 ; board $7.00 to $10.00 per week ; $2.00 per day. Pyramid Lake abounds in trout and bass, and is held exclusively for use of guests ; the best deer, partridge and small game hunting ground in the Adirondacks.

Mrs. A. R. Russell, Russell Cottage ; Private Boarding ; can accommodate 30 ; terms, $8.00 to 12.00 per week ; full particulars on application.

NORTH RIVER, P. O. WARREN CO., N. Y. (North Creek Station.)

Harrison Roblee, formerly Eldridge's, North River Hotel ; 4½ miles from North Creek Station ; accommodate 40 ; rates, $10.00 to $12.00 per week ; $2.00 per day.

HOTELS AND BOARDING HOUSES—*Continued*.

INDIAN LAKE, P. O. HAMILTON CO., N. Y. (North Creek Station.)

Frank Moody, Indian River Hotel; 16 miles from North Creek Station, on road to Blue Mountain Lake; accommodate 35; rates, $10.00 to $12.00 per week; $2.00 per day.

P. Ordway, Ordway House; 18 miles from North Creek Station, on road to Blue Mountain Lake; accommodate 60; rates, $10.00 to $12.00 per week; $2.00 per day.

John St. Marie, Boarding House; 18 miles from North Creek Station, on road to Blue Mountain Lake; accommodate 20; rates, $8.00 to $10.00 per week; $1.75 per day.

Alonzo Cole, Cedar River House; 20 miles from North Creek Station, on road to Blue Mountain Lake; accommodate 35; rates, $10.00 to $12.00 per week; $2.00 per day.

MINERVA, P. O. ESSEX CO., N. Y. (North Creek Station.)

William L. Keyes, Vanderwarker House on banks Vanderwarker Creek, one of the best trout streams in Adirondacks; 20 miles from North Creek Station; accommodate 10; rates, $10.00 to $12.00 per week; $2.00 per day.

James Doherty, Aiden Lair Lodge; 16 miles from North Creek Station; accommodate 15; rates, $8.00 to $10.00 per week; $2.00 per day.

Orson Kellogg, Minerva; 8 miles from North Creek Station; accommodate 20; rates, $8.00 to $10.00 per week; $1.75 per day.

OLMSTEADVILLE, P. O. ESSEX CO., N. Y. (North Creek Station.)

Patrick Sullivan, Alpine Hotel; 6 miles from North Creek Station; daily mail; accommodate 35; rates, $10.00 to $12.00 per week; $2.00 per day.

NEWCOMB, P. O. NEWCOMB, ESSEX CO., N. Y. (29 miles from North Creek Station.)

Harrison H. Williams, Newcomb Hotel; accommodate 35; rates, $8.00 to $10.00 per week; $1.75 per day.

James Hall, Half-Way House; accommodate 15; rates, $7.00 to $8.00 per week; $1.25 per day.

Washington Chase; accommodate 10; rates, $7.00 per week; $1.25 per day.

I. O. Bradley, Central House; accommodate 15; rates, $7.00 to $8.00 per week; $1.25 per day.

LONG LAKE, P. O. LONG LAKE, HAMILTON CO., N. Y. (39 miles from North Creek Station.)

E. Butler, The Sagamore; new, elegantly furnished; 3 miles from head of Long Lake; 9 miles from Blue Mountain Lake by road; accommodate 150; rates, $12.00 to $20.00 per week; $3.00 per day.

Mrs. C. H. Kellogg, Lake House; accommodate 50; rates, $10.50 per week; $2.00 per day.

HOTELS AND BOARDING HOUSES—*Continued.*

LONG LAKE, P. O. LONG LAKE, HAMILTON CO., N. Y. (39 miles from North Creek Station.)—*Continued.*

David Helms, Grove House ; accommodate 20 ; rates, $8.00 to $10.00 per week ; $2.00 per day.

Helms & Smith, Long Lake House ; accommodate 20 ; rates, $10.00 to $12.00 per week ; $2.00 per day.

Henry Austin, Austin's Cottage : accommodate 15 ; rates, $7.00 to $10.00 per week ; $1.75 per day.

William Kellogg, Island House ; accommodate 15 ; rates, $10.50 per week ; $2.00 per day.

BLUE MOUNTAIN LAKE, P. O. BLUE MOUNTAIN LAKE, HAMILTON CO., N. Y. (North Creek Station.)

G. W. Tunnicliff, Manager, Prospect House ; hotel 30 miles from North Creek ; by stage daily (except Sunday) ; accommodate 500 ; terms $20.00 per week and upward ; $3.50 per day. Blue Mountain and Raquette Lake steamer start from dock at foot of lawn daily except Sunday.

John Holland, Blue Mountain Lake House ; hotel, this new house, rebuilt 1887, and cottages connected with it, accommodate 250 ; rates, $15.00 to $20.00 per week ; $2.00 per day. Blue Mountain and Raquette Lake steamer start from dock at foot of lawn daily except Sunday.

Tyler M. Merwin, Blue Mountain House ; 200 feet above Blue Mountain Lake, and affords one of the finest views in the Adirondacks ; stages daily to and from North Creek ; capacity 50 ; rates, $10.00 to $15.00 per week ; $2.50 per day.

Edward Bennett, Under the Hemlocks, Raquette Lake ; P. O. Blue Mountain Lake. Under the Hemlocks is the suggestive title of Bennett's Hotel at Long Point. In connection with cottages, has accommodation for 90 ; rates, $10.00 to $20.00 per week.

Isaac Kenwill, Raquette Lake House, on Raquette Lake ; capacity 60 ; $12.00 per week ; $2.50 per day.

C. W. Blanchard, Blanchard's Wigwams on Raquette Lake ; capacity 20 ; rates, $10.50 per week ; $2.00 per day.

Chauncey Hathorns, Forest Cottages and Camps on Raquette Lake ; fine hunting and fishing in the immediate vicinity ; terms, $10.50 per week ; $1.50 per day.

Charles H. Bennett, The Antlers ; Raquette Lake ; an elegantly fitted new cottage and a number of rustic lodges, and dining and kitchen accommodation for 75 persons.

Joseph Whitney, Whitney's Camp ; accommodate 10 ; rates, $10.50 per week ; $1.50 per day.

Miron Fletcher, Forked Lake House, on Forked Lake ; accommodate 40 ; rates, $10.00 to $12.00 per week ; $2.00 per day.

Charles J. Griffin, Eagle Nest Cottage, located on north shore of Eagle Lake, formerly home of "Ned Buntline ;" accommodate 12 ; rates, $2.00 per day.

Samuel Davis, Forest House ; 25 miles from North Creek Station, on road to Blue Mountain Lake ; accommodate 10 ; rates, $7.00 to $10.00 per week ; $1.75 per day.

73

REACHED BY THE ADIRONDACK RAILWAY VIA SARATOGA, AND NORTH CREEK.

FORTY MILES BY LAKE AND STREAM!

One-day Excursion on elegant steam launches of Blue Mountain and Raquette Lake
Steamboat Line, through Blue Mountain Lake, Eagle Lake,
Utowana Lake, Marion River and Raquette Lake
to Forked Lake Carry and return.

FOUR LAKES, THREE INLETS, AND THE MARION RIVER.

LOW EXCURSION RATES ON REGULAR DAILY BOATS.

ALSO

Special Steam Yachts for Charter.

Apply to HENRY BRADLEY, SUPT., Blue Mountain Lake, Hamilton Co., N. Y.

"The D&H"

THE DELAWARE & HUDSON R. R.

"THE FAVORITE TOURIST ROUTE" TO THE GREAT

Adirondack Mountains,

Lake George, Lake Champlain, Ausable Chasm, Cooperstown, Sharon Springs and the Gravity Railroad.

EXCURSION TICKETS to the above-named points are on sale at the Albany and Saratoga Offices of the Company at low rates of fare.

The shortest and most comfortable route between NEW YORK and MONTREAL.

The "D & H" is the only line running Pullman Palace Sleeping Cars between **Albany** and **Chicago**. Passengers to and from the West by this route, in connection with the New York, Lake Erie & Western R. R. via Binghamton, pass through some of the most charming scenery in the country.

TRACK, EQUIPMENT AND SERVICE UNSURPASSED.

For maps, time-tables and descriptive guide, address,

J. W. BURDICK, General Passenger Agent,

H. G. Young, *Ass't Pres. and Gen'l Manager.*

Delaware & Hudson Canal Co.'s Railroad, Albany, N. Y.

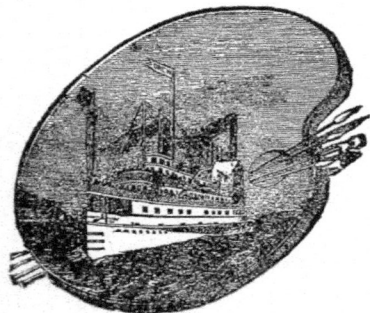

77

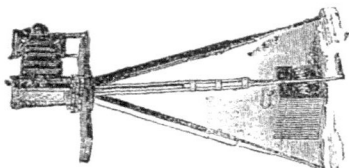

THE "WAYSIDE"

LAKE LUZERNE, N. Y.

700 FEET ABOVE TIDE WATER.

WESTERN UNION TELEGRAPH IN THE HOUSE.

One Hour's ride from Saratoga. At the confluence of the Sacandaga and Hudson Rivers.

Ten (10) Furnished Cottages on the Grounds. Fine Lawns, Shade and Walks.

H. J. ROCKWELL, Manager.

GEORGE W. FERRIS,

PROPRIETOR.

—+—
+

Board $10.50 to $12.00 per week; $2.00 per day. Accommodations for 150 guests. Open all the year. High, dry land; pure air; pure water from mountain spring; sanitary conditions perfect. Near to the best of fishing and hunting for birds and small game. 'Bus to Friend's Lake, 2 miles; Loon Lake, 2 miles; Brant Lake, 5 miles, or to Schroon Lake, for fishing parties, at 25 cents each. Livery connected with hotel.

THIS Hotel is located among the many spurs of the lower Adirondacks, being fifteen hundred feet above sea level, and within short drives of all the principal lakes, mountains and brooks, for which this region is so famous. Fine tennis and croquet grounds are in close proximity to the house. A daily mail from each direction is received. Direct telegraphic communications and two trains in each direction daily, which makes it very convenient for business men wishing to leave the city on Saturday and spend Sunday with their families in the country. All New York papers received at 2 P.M. Address:
GEO. W. FERRIS,
CHESTERTOWN, N.Y.
—+—
+
FISHING and BOATING.

Chestertown, - CHESTER HOUSE, - New York.

81

Taylor * House,

AND 12 COTTAGES

IN CONNECTION.

CAPACITY FOR 125 GUESTS.

Lake View Point.

P. O. SOUTH SCHROON, ESSEX CO., N. Y.

OPEN JUNE 1st, 1887.

—o—

❖TERMS❖

$2.50 to $3.00 per day, $12.00 to $15.00 per week.

Special arrangements made for families and a prolonged stay.

For particulars address

C. F. TAYLOR & SON,

Proprietors.

—o—

See description on another page.

Leland House and Cottages, Schroon Lake, Essex Co., N.Y.

L. R. & E. D. LOCKE.

Rates, $12.50 to $21.00.   Transient, $3.00 to $3 50 per Day.

For the Season of 1887.—The Leland House will be open for the reception of guests from June 10th to October 1st, under the proprietorship of L. R. & E. D. LOCKE, both well-known hotel men.

A new reading-room and bathrooms have been added since last season, and important improvements made in the office and parlors that will add greatly to the comfort and convenience of guests.

The Leland House is situated on a point at the head of Schroon Lake, and commands a view of the lake from three sides, the neighboring mountains, and the beautiful scenery of the Schroon Valley. It has a frontage of 162 feet facing south, across which and the eastern end of the building extends a grand piazza 22 feet high, 13 feet wide, and 256 feet long. The observatory, 107 feet above the lake. affords a prospect equalled by few in the Adirondacks. The grounds, about five acres in extent, are tastefully laid out, and extend to the lake on the south and east. This house was established in 1872, and at once took front rank as a summer hotel.

The sanitary provisions are very complete. The main office contains telegraph and news stand. Capacity of house and cottages about 200.

PLAN OF PARLOR FLOOR,

PROSPECT HOUSE,

BLUE MOUNTAIN LAKE,

Hamilton Co., N. Y.

——o——

The largest and best hotel in the Adirondacks, 2,000 feet above the sea level, and said to be the most complete mountain house in the United States. The building is plain and unostentatious while massive and substantial. The entire house is illuminated with the Edison incandescent light. A hydraulic elevator renders all of its upper floors equally accessible. The

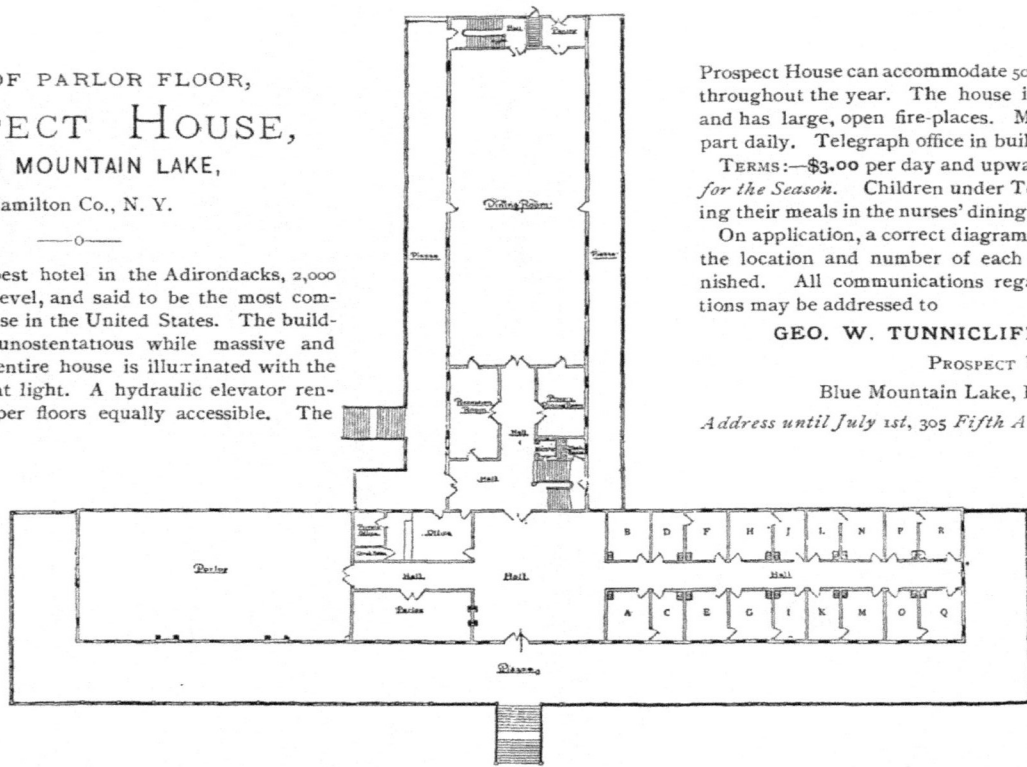

Prospect House can accommodate 500 guests and is open throughout the year. The house is heated by steam, and has large, open fire-places. Mails arrive and depart daily. Telegraph office in building.

TERMS:—$3.00 per day and upwards. *Special rates for the Season.* Children under Ten years of age taking their meals in the nurses' dining room—half price.

On application, a correct diagram of the floors, giving the location and number of each room, will be furnished. All communications regarding accommodations may be addressed to

GEO. W. TUNNICLIFF, Manager,

PROSPECT HOUSE,

Blue Mountain Lake, Hamilton Co., N. Y.

Address until July 1st, 305 Fifth Avenue.

Dining Room.

Piazza. Piazza.

Parlor

Office

Hall Hall

Parlor

Piazza

B D F H J L N P R

A C E G I K M O Q

Hall

84

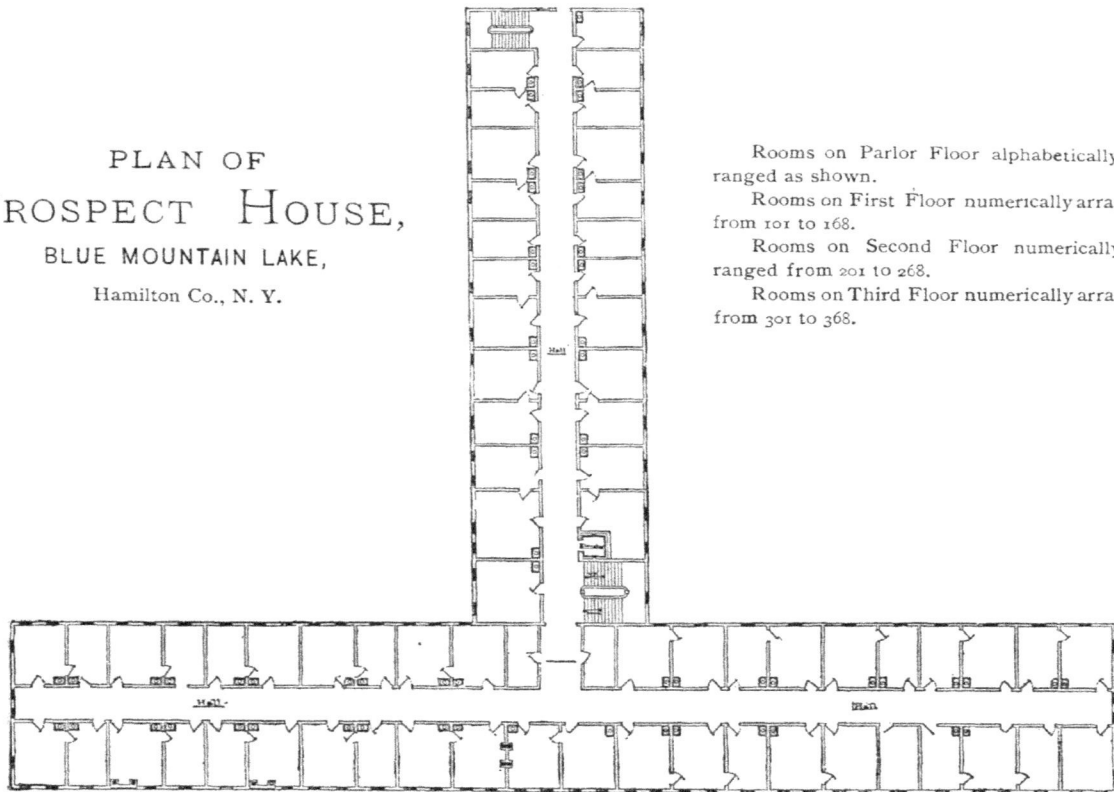

PLAN OF
PROSPECT HOUSE,
BLUE MOUNTAIN LAKE,
Hamilton Co., N. Y.

Rooms on Parlor Floor alphabetically arranged as shown.

Rooms on First Floor numerically arranged from 101 to 168.

Rooms on Second Floor numerically arranged from 201 to 268.

Rooms on Third Floor numerically arranged from 301 to 368.

Hall

Murray·Hill·Hotel

PARK AVENUE 40TH AND 41ST STS,
NEW YORK.
HUNTING AND HAMMOND.
86

THE MURRAY HILL HOTEL,

PARK AVENUE, 40th & 41st Sts.,

Is the most convenient of any hotel in New York City for travelers or tourists, being only ONE MINUTE'S walk from Grand Central Depot.

NO CARRIAGE HIRE,

NO CHARGE FOR BAGGAGE,

As we transfer it to and from Grand Central Depot free.

It is the only FIRST-CLASS hotel in New York City. On both American and European plans.

HUNTING & HAMMOND.

87

OFFICE, CORRIDOR & MAIN STAIRWAY.

MURRAY HILL HOTEL.

90

OVER THE CARRY.

J. H. RUSHTON,

CANTON, N.Y.,

BUILDS

PLEASURE BOATS,

HUNTING BOATS, SNEAK BOATS, SAILING AND

PADDLING CANOES, STEAM LAUNCHES,

to order, and has in stock

OARS, ROWLOCKS, SAILS, CLEATS, BLOCKS, ETC., ETC.

FINE GOODS A SPECIALTY.

Send FIVE Cents for 80 page Illustrated Catalogue.

92

THE ONTARIO CANOE CO.,

(LIMITED.)

PETERBOROUGH, - - - - - ONTARIO, CAN.

PLEASURE, FISHING AND HUNTING

CANOES

(GOLD MEDAL, LONDON, 1883.)

SEND 3 CENT STAMP FOR ILLUSTRATED CATALOGUE.

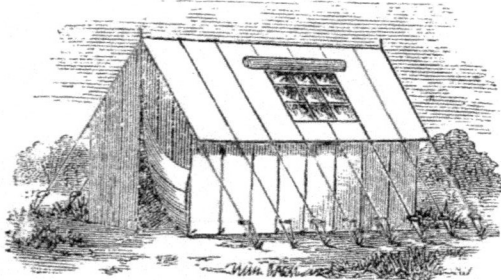

Manufacturer of Tents of all kinds, sizes and shapes,
at low figures.

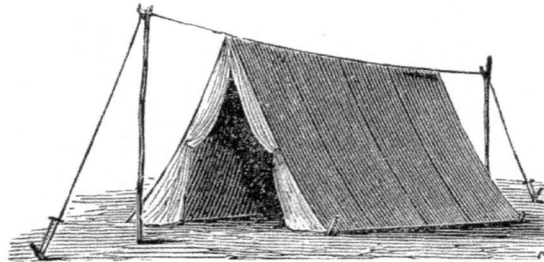

A Tent or Common Tent, with rope ridge, used to
camp out with, made any size and of best
goods.   Prices from $6 to $12.

Window & Store Awnings.

Canopy for Croquet Grounds, Sea Shore, Gardens or Lawn Tennis,
made in the best manner and of best goods.

Flags and Burgees of all
kinds made to order.

All these goods made in the best manner and very lowest figures. All warranted mildew and water-proof at a cost of
3 cts. per yard extra.

S. HEMMENWAY, 60 South Street, New York.

THE UNEXCELLED FIREWORKS COMPANY,

(INCORPORATED 1874.)

9 & 11 PARK PLACE, N. Y.

Wes'ern House: 519 LOCUST STREET, St. Louis, Mo.

————o————

LARGEST MANUFACTURERS AND LEADING HOUSE IN

FIREWORKS, FLAGS, LANTERNS, BALLOONS, &c.

————o————

The four largest, finest, most novel and best managed

DISPLAY OF FIREWORKS

ever produced were those at

THE CENTENNIAL AT NEWBURGH, N. Y.,

OCTOBER 18th, 1883.

THE PRESIDENTIAL INAUGURATION,

AT WASHINGTON, MARCH 4th, 1886.

THE BI-CENTENNIAL AT ALBANY, N. Y.,

JULY 22d, 1886, and

THE UNVEILING OF THE BARTHOLDI STATUE OF LIBERTY at New York City, Nov. 1st, '86.

They were manufactured and furnished by

THE UNEXCELLED FIREWORKS CO., 9 & 11 Park Place, N. Y.

☞ SEND FOR ILLUSTRATED CATALOGUE—FREE.

A Great Success!   Works Like a Charm!!   Saves Time and Eyesight!!!

SIMPLICITY.   DURABILITY.   UTILITY.

ACME BUTTON-HOLE ATTACHMENT,

FOR GENERAL USE.

Can be Readily Attached to any Family Sewing Machine. Will make three button-holes a minute, PERFECT IN FINISH—OF ANY SIZE—on goods of any texture. Being automatic, any person can operate it. Cannot get out of order. Gives perfect satisfaction. This is the ONLY MACHINE ATTACHMENT IN THE WORLD that will make a button-hole superior in finish and durability to the best hand-made button-hole. **Price, $5.00.**

For descriptive circulars and samples of work address

ACME BUTTON-HOLE ATTACHMENT CO. (LIMITED),

MAIN OFFICE AND SALESROOM:

74 Fifth Avenue, near 14th Street, New York.

97

THE LIGHT RUNNING

FINEST WOODWORK.   EASIEST TO MANAGE.

NEW   HOME

HAS NO EQUAL.   SIMPLEST AND BEST.

SAN FRANCISCO, Cal.   SEWING MACHINE.   ATLANTA, Georgia.
     ST. LOUIS, Mo.   DALLAS, Texas.

30 Union Sq.,   New Home Sewing Machine Co.,   248 STATE St.,
   New York.   Chicago.
              ORANGE, MASSACHUSETTS.

PRICE, $85.00.

CHARLES DALY—THREE BARREL GUN

is the best weapon yet produced for large and small game.

QUALITY SUPERB. 12 GAUGE SHOT—32, 38, 40, 45 RIFLE.

CHARLES DALY—HAMMERLESS GUN, PIEPER GUNS, MANHATTAN GUNS, BALLARD RIFLES.

ASK YOUR DEALERS FOR THEM.

84 & 86 CHAMBERS ST.,   -   Agents: SCHOVERLING, DALY & GALES,   -   *NEW YORK.*

99

In or Out of the ADIRONDACKS

via the ADIRONDACK RAILWAY.

TO OGDENSBURG & 1000 ISLANDS

www.ingramcontent.com/pod-product-compliance
Lightning Source LLC
Chambersburg PA
CBHW021943190326
41519CB00009B/1122